DNA and Genetics

Anna Kessling

science
museum

For Lydia, Melanie, Thomas, Ewan and all our relatives.

Author's acknowledgements

Dr Gail Davies, Dr Una Fairbrother and Ewan Smith for helpful comments on the text, and Lawrence Ahlemeyer for forbearance.

Published 2003 by NMSI Trading Ltd,
Science Museum, Exhibition Road, London SW7 2DD.

British Library Cataloguing-in-Publication Data
A catalogue record for this publication is available from the British Library

Designed by Jerry Fowler
Printed in Belgium by Snoeck-Ducaju & Zoon

Additional photography by David Exton, Claire Richardson and Jennie Hills

ISBN 1 900747 55 3

Website http://www.nmsi.ac.uk

Contents

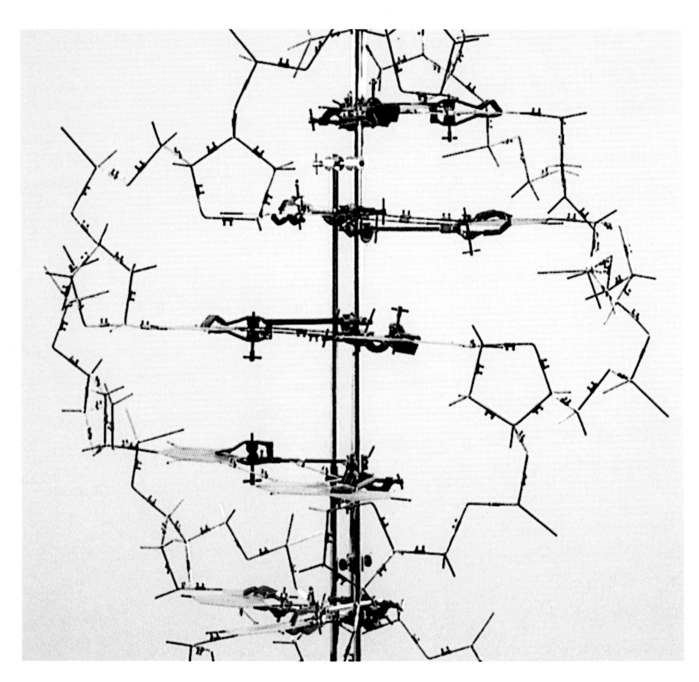

4

Introduction

This book introduces the basic concepts of DNA and genetics, using the Science Museum's collections to illustrate this important area of modern science, technology and medicine.

In 1953, two scientists, James Watson and Francis Crick, worked out the structure of DNA: they realised it was a 'double helix' and built a model to show this.
A reconstructed version of this model, which you can see opposite, is now in the Museum. The excitement of this breakthrough has not diminished with time. Half a century later, the technological advances which it made possible still affect all our lives.

Watson and Crick's discovery opened up many new areas of knowledge. Once scientists understood the elegant structure of DNA, they could start working out its roles – in the inheritance of genetic features and in the division of cells, for example. In turn, this knowledge paved the way for advances in biotechnology – such as genetic tests, genetically modified crops and transgenic animals – which bring with them both benefits and controversies.

The book also explores some aspects of DNA in more depth, to let you share the enthusiasm of a scientist working in genetics today.

Anna Kessling

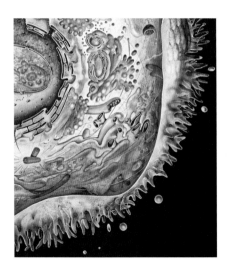

This is an artist's view of the contents of a cell. It gives you an idea of just how complicated and busy cells are. Each cell is a complex system in itself, and cells in different parts of an organism have specialised structures or shapes to enable them to carry out particular tasks. (Science & Society Picture Library)

DNA and genetics in brief

Cells

Apart from the simplest types of living organisms, which may have only one cell, animals and plants are made up of many cells.

There are different types of cells in the different parts of living organisms. For example, it would not be much use to us if we had toenail-making cells where our brain cells should be, or skin cells where we need to have bone cells. Roses need to have petals above ground and roots below; a goldfish needs its fins in particular places on its outside or it would not be able to swim properly.

Cells contain a specialised area called the nucleus. In humans the only cells without a nucleus are the red blood cells.

DNA

The nucleus of the cell contains the genetic material – a substance called DNA (deoxyribonucleic acid). The DNA provides the essential instructions for the building, maintenance, reproduction and repair of the organism.

DNA is contained in structures called chromosomes, which are like very fine threads. The chromosomes in each cell contain the complete DNA instructions for that organism.

Chromosomes

In humans, each cell with a nucleus has 46 chromosomes (eggs and sperm have only 23 chromosomes each). The 46 chromosomes consist of 22 pairs plus an extra pair (known as the sex chromosomes). Other organisms have different numbers of chromosomes: the fruit fly, *Drosophila*, has eight chromosomes (three matching pairs and a pair of sex chromosomes). Human chromosome pairs are numbered from 1 to 22 (the 23rd pair are the sex chromosomes).

In humans and other diploid organisms (those with two matching sets of chromosomes) there are two sets of DNA instructions in each cell. The DNA instructions at a particular point on one chromosome of a pair match those at the same point on the other chromosome of the pair.

The 23 pairs of human chromosomes, identified under a microscope in a dividing cell, and rearranged so they are easier to compare. This kind of image is called a karyotype. We can tell that this particular karyotype comes from a woman. Can you see why? During the nineteenth century, scientists showed that the nucleus contains the genetic information. By 1903 there were good theories in favour of the idea that the chromosomes transmit genetic information. Early-twentieth-century scientists knew that chromosomes contained DNA, and they thought it must be the genetic material. (Image by permission of MetaSystems GmbH)

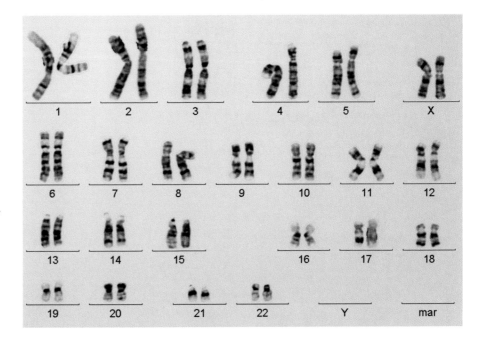

In humans, each of the 22 pairs (neatly) contains one chromosome inherited from the mother and one from the father. This means that parents each pass half their own genetic information to each child.

Sex chromosomes

What about the sex chromosomes in humans? Sex chromosomes are called X and Y chromosomes. In humans (and many other species), females have a pair of X chromosomes – one inherited from the mother and one from the father. Males have an X chromosome and a Y chromosome – the X chromosome inherited from the mother and the Y chromosome inherited from the father.

So every human baby inherits an X chromosome from its mother – but a boy also inherits a Y chromosome from his father, while a girl inherits an X chromosome from her father as well as one from her mother.

Each sperm made by a man's body contains one chromosome from each of his 22 pairs, and either an X chromosome or a Y chromosome. Each egg made in a woman's body contains one chromosome from each of her 22 pairs and an X chromosome.

DNA and genetics in brief

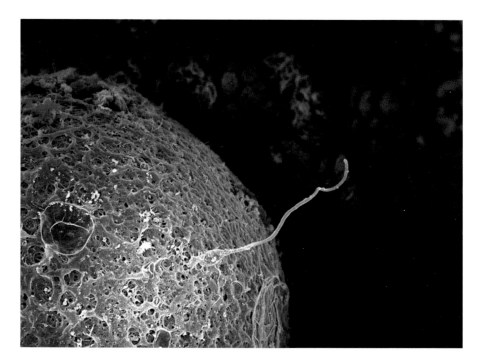

Zygotes

A particular combination of a woman's egg with a man's sperm makes a fertilised egg or zygote (pronounced *ZY-goat*). The egg and sperm join together to make a new cell which contains 22 pairs of chromosomes and two sex chromosomes. Each parent has supplied one chromosome of each pair and one sex chromosome. The zygote contains all the genetic information needed to make a new person.

Other diploid organisms have different numbers of chromosomes. In organisms that have two sexes, the processes that reduce the chromosome number in the gametes (equivalents of the egg and sperm in humans) and that restore the full number of chromosomes in the zygote both happen in similar ways.

1 The structure of DNA

1.1 Genetics in 1953

If we could travel back in time to 1953, we would already find many of the modern inventions and discoveries we now take for granted. Cars, televisions, radios, aeroplanes, aspirin, antibiotics – all of these were available in 1953. By contrast, today's advances in genetics would have seemed fabulous to the scientists of 1953.

At that time, scientists thought that the genetic material in humans was contained in 48 chromosomes (the thread-like structures found in the nucleus of most of the body's cells). They also knew that the chromosomes contained DNA (deoxyribonucleic acid), and they were convinced that DNA must be the material that controlled inheritance. But nobody was quite sure how DNA was put together, or how it worked.

As often seems to happen with scientific advances, there was a rapid accumulation of knowledge in the few years leading up to the major breakthrough – in this case the breakthrough was the publication of the structure of DNA. Three key discoveries set the scene.

1.2 Two earlier discoveries

First, in 1950, the American scientist Erwin Chargaff (1905–2002) showed that two of the chemical components of DNA are present in exactly the same proportion as two other components – irrespective of the origin of the DNA. We now refer to these components as 'DNA bases' (see page 13). Chargaff showed that the proportion of guanine (G) always equals the proportion of cytosine (C), and that the proportion of thymine (T) always equals the proportion of adenine (A). This was an interesting finding, but was it any more than a quirk of nature? In fact, it was to prove crucial to the understanding of the DNA structure.

The next key discovery came in 1952, when two more American scientists, Alfred Hershey (1908–97) and Martha Chase (b. 1930), showed that in a particular type of virus which, conveniently, consisted only of DNA and protein, the DNA contained all the genetic information. This confirmed scientists' hunch that DNA was the genetic material.

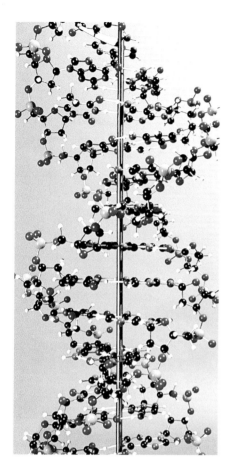

Figure 1.1 Watson and Crick (in collaboration with Rosalind Franklin and Maurice Wilkins) worked out the structure of DNA, the molecule that holds the genetic instructions for the construction, running and maintenance of all living organisms, except for a few viruses. This is a space-filling model of the structure of the DNA molecule, with the sizes and relationships of the components shown in huge magnification, but in realistic proportions to each other. (Science & Society Picture Library)

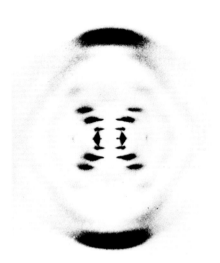

Figure 1.2 An image of the structure of DNA, produced using the same technique Rosalind Franklin used when she worked out that the DNA molecule must be shaped like a coil. What is special about this odd-looking pattern? We can gain a very basic understanding of the issue which faced Rosalind Franklin if we take the curly cable of an appliance like a telephone (remember to unplug anything electrical before you try this with its cable). Hold the cable lengthways so its shadow falls along a plain, flat surface. At a precise distance from the surface, the shadow of the cable is in sufficiently sharp focus that we can tell, from the shadow alone, that the cable is probably curly – or helical. However, if we hold the cable at a right angle to the surface, and shine the light along the length of the cable instead of through it, we can see that it would be much harder to guess the structure of the cable from the resulting shadow. (Science Photo Library)

1.3 Rosalind Franklin

The British crystallographer Rosalind Franklin (1920–58), working with Maurice Wilkins, found the third essential clue to the structure of DNA. She had managed to obtain high-quality DNA in a fibre form, and she studied the structure of the fibres using an approach that is still practised today, for instance in determining protein structure. Franklin passed a beam of X-rays through the DNA fibre, end-on, and studied the resulting pattern made by the beam. Others had not succeeded in doing this. Franklin's results showed an X-ray pattern which strongly suggested that there was something helical about the structure of DNA.

Her study of these patterns was equivalent to interpreting the shadow cast after a light had been shone the length of a structure like Figure 1.1, and working out from that shadow (Figure 1.2) that the shape must be a helix. She made a major scientific advance.

1.4 Watson and Crick

Although the X-ray diffraction pattern that Franklin produced suggested a helical shape, this did not fit with what was known about the composition and size of DNA. A single helix could not explain the findings. Then, in 1953, two scientists working at Cambridge University, James Watson (b. 1928), who was American, and Francis Crick (b. 1916), who was British, made an inspired and revolutionary discovery. They worked out that DNA must be a double helix – two coils linked together like a twisted ladder (Figures 1.3 and 1.4).

1.5 The double helix

DNA is described as a 'right-handed' double helix. Everyday examples of helices can be found in curly telephone cables, as we have seen, or spiral staircases. What makes a helix 'right-handed'? You can begin to make a right-handed helix with your right arm. Hold out your right forearm, with the palm facing up. Bend your arm at the elbow so your fingers point towards the ceiling. Now keep your fingers together and twist your hand so your little finger moves towards your face, your thumb moves away from your face, and your other fingers curve round, their tips still pointing to the ceiling. Your forearm and hand now make a right-handed helix. If you do exactly the same with your left forearm and hand, you will make a left-handed helix, the mirror image of a right-handed helix. Try looking at a telephone cable and seeing if you can tell whether it is left- or right-handed. Sometimes you can find both left- and right-handed segments in the same cable.

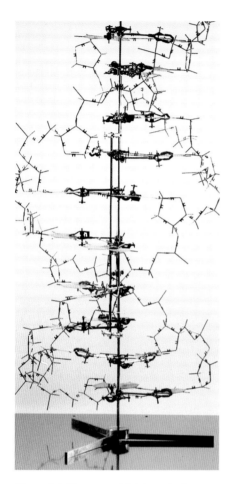

Figure 1.3 Watson and Crick worked out that there must be exactly ten base pairs for each complete turn of the DNA double helix. This is a reconstruction of their original model of one complete turn of the molecule, using many original parts. The flat plates represent the DNA bases and the spiky structures at the outside of the model are the helical backbone. A close-up view is shown on the front cover and in Figure 1.6. (Science & Society Picture Library)

Figure 1.4 James Watson and the DNA model at the Science Museum in 1994. Watson and Crick's original paper about their discovery appeared in the prestigious scientific journal Nature. It occupied less than one whole page of the journal – yet it reported the discovery that is the whole foundation of our current understanding of genetics. Without their work we would not have molecular biology, genetic engineering, cloning, gene therapy or the human genome project. Neither would we be able to diagnose single-gene defects before birth, or identify the contribution of genes to the risks of getting common diseases such as cancer. (Science & Society Picture Library)

The structure of DNA

Figure 1.5 The structure of the DNA helix. The left-hand drawing shows the helix as a straight ladder, so you can see an example arrangement of base pairs. The right-hand drawing shows the ladder's actual shape, twisted into a helix. In reality the base pairs lie flat, like the rungs of the ladder, but here they are shown upright so they are easier to see. Figure 1.6 shows a similar series of base pairs in Watson and Crick's model, starting from the CG pair at the top.

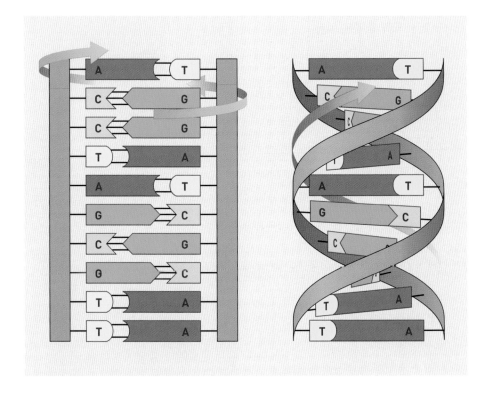

To understand what the double helix of DNA itself looks like (Figure 1.5), imagine it as a long, straight ladder with flexible uprights and ten solid, flat, horizontal rungs. Facing the ladder, keep the base still, and take the top of each upright in one hand. Then twist the uprights around, pushing the right upright away from you, and pulling the left upright towards you. Carry on twisting until the tops of the uprights are back to where they started, having travelled through 360 degrees. The rungs remain parallel to the ground. Now you have a rather large outline model of one complete turn of the DNA double helix (one helix from each upright) with ten base pairs. DNA goes on for many thousands of base pairs in this way.

The 'uprights' in DNA are called its backbone. They consist of a regular arrangement of a type of sugar, called deoxyribose (the 'D' of DNA), with small chemical phosphate groups (the 'sugar–phosphate backbone').

In a straight ladder, the supporting uprights have to be kept the same distance apart. In DNA, the uprights are the same distance apart because each rung is made from a

Figure 1.6 *This is a closer view of part of Watson and Crick's model of DNA, showing base pairs in close-up. We can read off the pairs of bases from top to bottom and from left to right in this picture as CG, CG, TA, AT, and so on. (Science & Society Picture Library)*

pair of particular chemicals called bases. One base of each pair is a purine (the purines which go into DNA are guanine and adenine); the other base of each pair is a pyrimidine (the pyrimidines which go into DNA are thymine and cytosine). The purines are bigger than the pyrimidines, but the two purines are about the same size as each other, and the two pyrimidines are also about the same size as each other. Since all the 'rungs' or base pairs have to contain one purine and one pyrimidine, they will all be about the same size. In fact, there is a further restriction on how the structure can vary, because only two combinations of bases are permitted. Thymine pairs with adenine (T with A); guanine pairs with cytosine (G with C). The pairing is through weak chemical bonds called hydrogen bonds, three to join a G to a C and two to join an A to a T. Just as the rungs of our imaginary ladder lie parallel to the floor, so the base pairs 'sit' flat, at right angles to the long axis of the DNA. There are ten base pairs per complete DNA turn. Figure 1.6 shows how this was represented in Watson and Crick's model.

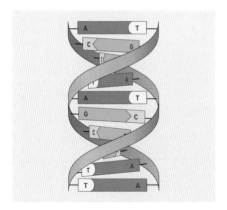

Figure 1.7 The structure of the DNA helix.

1.6 Pairing is what DNA bases do best

The DNA bases are 'made for each other' – they align and pair with each other, because their shapes make them fit together in a particular way. If a piece of DNA double helix (Figure 1.7) is unwound and the two strands separate, then, even in a mixture of other single-stranded DNA samples, the two single strands that go together (complementary strands) are likely to 'find' each other and re-form the double helix. If a single strand of DNA is put into a test tube containing the right chemical building blocks, biological tools and environmental conditions, the matching or complementary strand will be made, re-forming the double helix. Being a double helix is what DNA does best. This precise pairing tendency has been exploited extensively both in nature (see DNA replication, pages 20–22) and in science.

So what does a sample of DNA, say, in a test tube, actually look like? The process scientists use to recover DNA from human blood is outlined in the separate panel (Figures 1.8–1.12).

Preparing DNA

Figure 1.8 This scientist is pouring a blood sample into a tube, so that DNA can be recovered from the blood. Chemicals are used to break the thin membrane surrounding the white blood cells, which lets most of the cell contents out, leaving the nucleus, which contains the DNA in the chromosomes. Then a combination of a detergent and a special enzyme is used to break the nuclear membrane and release the DNA from the chromosomes. (Science & Society Picture Library)

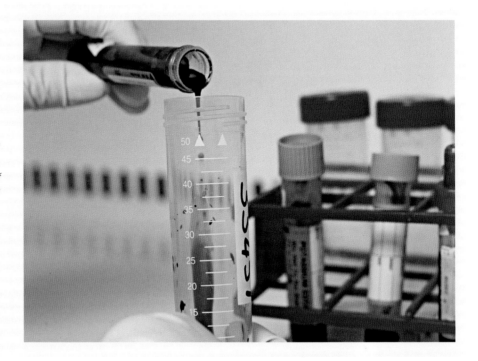

Figure 1.9 The tube then contains the DNA, plus the broken-down contents of the nucleus. These other substances can be removed by a combination of chemical treatment and centrifuging, to leave a clear solution. (Science & Society Picture Library)

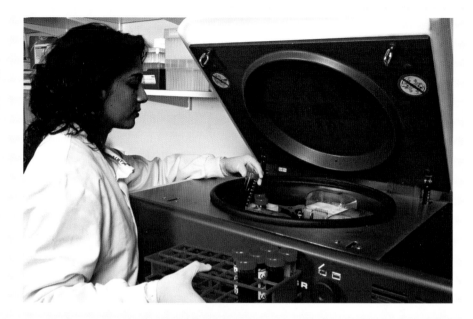

Figure 1.10 The final step of DNA preparation is to add some alcohol to the clear solution, to make the DNA insoluble. When the tube is given a gentle swirl, its contents seem to blur, and then threads of DNA form at the sides of the tube and merge in the middle. At this stage the DNA can be pulled out ('spooled') from the tube. The alternative to spooling DNA out of the solution is to leave it to come together in a creamy-coloured cloud, which comes to look, somewhat less poetically, like a glob of glue. (Science & Society Picture Library)

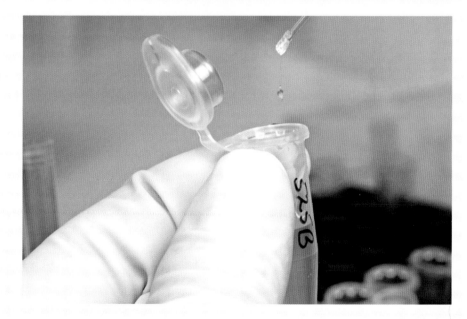

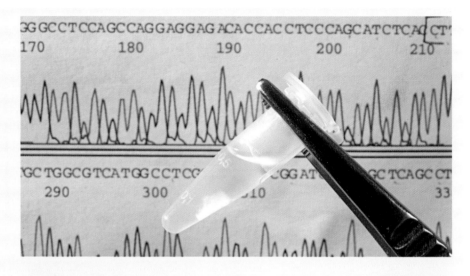

Figure 1.11 A DNA 'glob' in a tube. The background is an example of a DNA sequence chromatograph, which is produced when scientists carry out auto-mated DNA sequencing. DNA can be extracted from many sources. Examples include the root of a single hair, the centre of the tooth of an old skeleton, a spot of blood and the cells that come off on your finger if you rub the inside of your cheek. This means that forensic scientists may be able to recover DNA from surprisingly small samples at the scene of a crime. (Science & Society Picture Library)

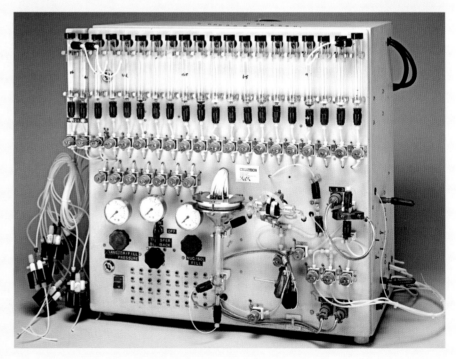

Figure 1.12 This is an early example, from about 1980, of a machine for synthesising short sequences of DNA. (Science & Society Picture Library)

2 How genes work

2.1 A store of genetic information

All the genetic information in DNA is contained in the particular sequence of the four bases in the helix. Think how amazing an idea this is. Within a particular species, the order and arrangement of the four DNA bases is essentially constant, with minor variations among individuals. Of course, the environment and non-genetic influences can contribute to physical differences between individuals and between species. However, small and large differences in the order and arrangement of the four DNA bases decide all the little (and big) biological differences between boys and girls, the reader and the author of this book, mice and elephants, spiders and cockroaches, aardvarks and zebras, roses and nettles, fish and fowl, dinosaurs and bacteria or, indeed, between any two living things (Figures 2.1 and 2.2) – leaving aside some viruses based on ribonucleic acid (RNA), which is itself related to DNA.

Figure 2.1 Tracy, a transgenic sheep, lived from 1990 to 1998. Because all species use their DNA in the same sort of way, even though their genes are different from each other, the cells of one species can be persuaded to use DNA from another species (a transgene) to produce a protein for the species from which the gene was originally taken. Tracy produced a human protein in her own cells because she had a human transgene. See pages 54–55. (Science & Society Picture Library)

Figure 2.2 Hair colour is an example of a physical difference controlled by genes. Several genes influence the amount and type of pigment in hair. (Science & Society Picture Library)

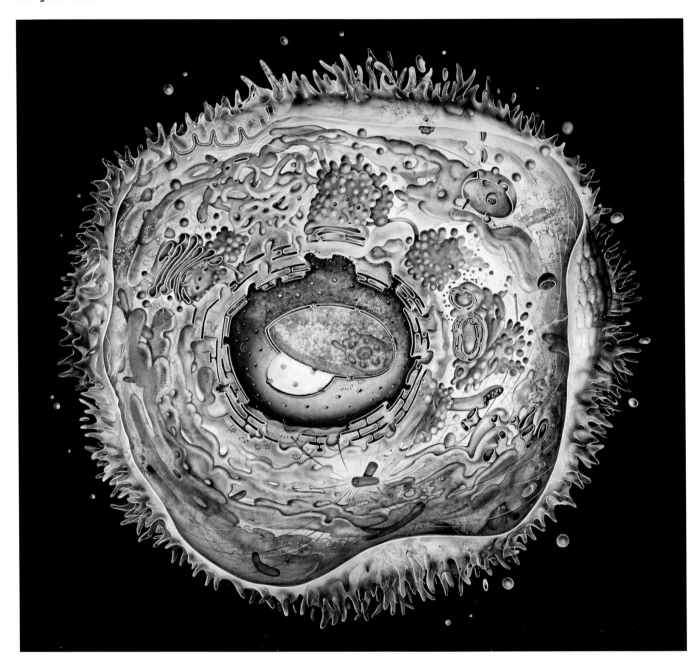

Figure 2.3 (opposite) A painting by John Barber and Cynthia Clark, showing the internal structure of an idealised animal cell. (Science & Society Picture Library)

2.2 Complete and detailed instructions

DNA contains the encoded instructions for the whole organism. So the complete instructions for making a plant are in the DNA of a single plant cell, the complete instructions for making a whole slug are in the DNA of a single slug cell – and the complete instructions for making a whole human are in a single human cell (Figure 2.3).

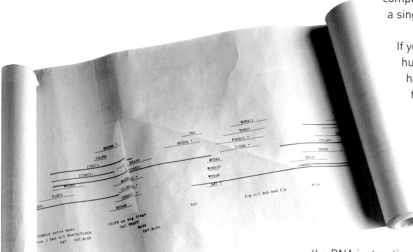

If you take the briefest look at a plant, a slug or a human, you will see that none of these organisms has the same consistency, texture or structure throughout (Figure 2.4). The plant has stem and leaf and may have a flower – yet all of these have the same DNA in each cell. The slug is longer than it is wide. It has feelers, a body and a flat foot underneath. Humans have many visible differences between parts of their bodies and they have many specialised internal organs – yet all of these have the same DNA in each cell. Why do all the parts of an organism not look the same? This is because the DNA instructions are extremely detailed and specific: what to make where, when to make it, when to stop making it, how much to make, and what else to make at the same time. These fine-tuning details for the instructions are encoded in what are called the regulatory sequences of DNA. The regulatory sequences control where and when a gene should be used to make proteins. And the genes contain the exact instructions on how to make these proteins.

Figure 2.4 Part of a 'chromosome map' of a nematode worm. Scientists produce maps like this to help them work out the complete arrangement of the genes needed to make a member of a particular species (the genome map of the species). (Science & Society Picture Library)

Examples of proteins include:

- albumin, which makes the white of a hen's egg and which is an important protein in human blood.
- blood-group determining proteins.
- collagen, which helps give strength to bones.
- enzymes, which help us digest our food and assist many of the body's chemical reactions at a subcellular level.

This characteristic, that all cells with nuclei contain all the DNA needed to make a whole organism, can be used in science for many applications – from investigating the normal production of a protein, to working towards gene therapy to treat and cure serious genetic disorders.

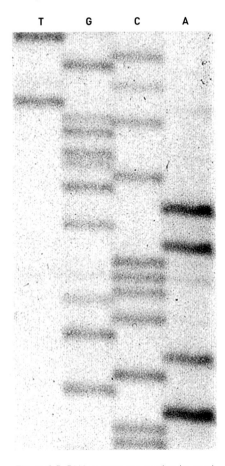

Figure 2.5 *DNA sequences used to be read from a sequence of bands, each of which represents a single base in the DNA. Individual columns in the black-and-white picture, which is called an autoradiograph, represent numbers of A, C, G and T bases. Scientists can 'read off' the sequence by taking into account the four columns referring to a particular piece of DNA and working up from the bottom of the autoradiograph, recording each base in turn as its band appears. (Una Fairbrother)*

2.3 Making proteins

The genes contain the complete set of instructions for making proteins, 'written' in the genetic code (see separate panel, Figures 2.5 and 2.6). So all genes are made of DNA, and the DNA in the genes encodes the proteins. However, some DNA sequences do not encode proteins. There are several examples, including the regulatory sequences we have already encountered.

The encoding of protein by DNA is not a direct mechanism. DNA is a large molecule in the chromosomes in the nucleus, and although proteins are manufactured in the cell, they are not made in the nucleus. So how does the DNA code reach the part of the cell that manufactures protein? First, the DNA sequence is transcribed (rewritten) into an RNA (ribonucleic acid) sequence. RNA is a molecule related to DNA, with three important differences: it is single-stranded, not double-stranded; it has a different chemical backbone; and its base sequence includes a pyrimidine called uracil (U) instead of thymine.

The whole gene sequence is 'read' from one strand of the DNA and rewritten or transcribed into RNA, without altering the code. The order of the bases in the DNA decides the order of the bases in the transcribed RNA, which we call messenger RNA or mRNA. This mRNA leaves the nucleus and moves into the outer part of the cell, the cytoplasm, and to specialised structures called ribosomes.

The order of the bases in the transcribed mRNA (rewritten from DNA) is translated from the 'language' of the DNA base sequence to the 'language' of the amino acid sequence that makes proteins. Special molecules called transfer RNA or tRNA carry out the translation. There is a tRNA to recognise the base sequence coding for each amino acid: each specific tRNA lines up its encoded amino acid in the order dictated by the message from the DNA sequence. The order of the bases in the transcribed RNA therefore decides the order of the amino acids in the protein.

So, the order of the bases in the DNA decides the order of the bases in the transcribed RNA. And the order of the bases in the transcribed RNA decides the order of the amino-acids in the protein. Thus the order of the bases in the DNA decides the order of the amino-acids in the protein encoded by the gene.

2.4 DNA replication

In nature, DNA's base-pairing properties enable it to make copies of itself – which allows cells to divide, to produce new cells that contain the same genetic information.

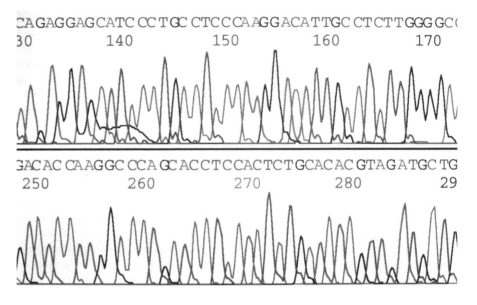

Figure 2.6 Modern DNA sequences are read in an automated way. First, the four different bases are labelled with dyes of different colours. The sequencing machine detects and analyses these dyes, providing a printout like this of the result. The top row of this sequence can be read directly from the sequencing software's output: CAGAGGAGCATCCCTGCCTCCCAAGGA, etc. (Science & Society Picture Library)

DNA replication has to happen accurately, precisely and at the right time for the particular type of cell. Cells divide to make new cells at different rates, depending on their purpose. For example, human skin cells keep dividing, so we can make new skin to replace the surface that rubs off. By contrast, the nerve cells in adult humans do

The genetic code

The genetic code was worked out some time after the discovery of the structure of DNA, by three scientists working in the United States in the 1960s: Marshall Warren Nirenberg (b. 1927), Har Gobind Khorana (1922–93) and Severo Ochoa (1905–93). They found that sets of three bases from the DNA sequence (codons) would be enough to encode the 20 amino-acids that are used, in turn, to build proteins. Then they worked out which particular set of three bases encodes which amino-acid.

With hindsight, it seems straightforward to realise that it might take three bases to encode one amino-acid. The reasoning is as follows. Twenty amino-acids are needed to make human proteins. A one-base-at-a-time code could only explain four

amino-acids, as there are only four bases. A two-bases-at-a-time code could explain up to 4x4, or 16, amino-acids – not quite enough. But a three-bases-at-a-time code could explain up to 4x4x4, or 64, amino-acids – which is more than enough. The first two bases of each set of three (each codon) are most critical in determining which amino-acid is encoded, so sometimes the third base of the codon does not affect which protein is made. For example, the mRNA codons which correspond to the DNA sequence triplets CGA, CGC, CGG and CGT all encode the amino-acid arginine. There are also specific codons that say 'start the protein here' and 'stop the protein here'. For instance, DNA triplets TAA and TAG both give a 'stop' instruction via mRNA, while TAT and TAC encode the amino-acid tyrosine.

not usually divide. When new cells are made, they need a new set of all the cell's contents, including an accurate copy of the DNA. In order to replicate, DNA has to unwind into its two constituent strands. Then, in essence, what happens is that new bases are lined up to pair with their 'partners' in the separated strands and the run of new bases alongside each separated strand is joined together to make a complementary strand for each of the original two strands. This process reconstitutes two double helices, each with the base sequence of the original double helix, each containing one strand from the original double helix and a newly made second strand. This is called semiconservative replication (see separate panel, Figure 2.7); the new DNA molecule is effectively half old and half new.

Figure 2.7 Semiconservative replication makes two double helices, each with the base sequence of the original helix. It works by a special mechanism that prevents the DNA chain from breaking.

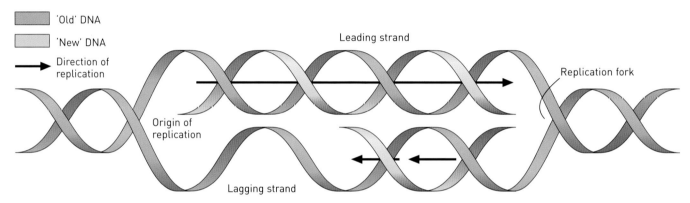

Legend:
- 'Old' DNA
- 'New' DNA
- → Direction of replication

Leading strand
Replication fork
Origin of replication
Lagging strand

Semiconservative replication

The direction of manufacture of the new DNA is dependent on the direction of the backbone. DNA starts replicating from a number of sites (origins of replication) along the chromosome. At these origins, the DNA is unwound a little, under the influence of enzymes. These enzymes act specifically to make the unwinding easier and to ensure the intact DNA to either side of the part that is replicating does not suffer excessive strain and break. The point at which the unwound DNA meets is called the replication fork – the replication fork moves along the DNA molecule as the DNA is replicated.

The enzymes that build up the new DNA molecules can only work in a particular direction, referred to as a 5' to 3' ('five prime' to 'three prime') direction. This terminology comes from a feature of the two DNA backbones, named after the numbering of the carbon atoms in the sugar molecules of the backbones. Each backbone in a pair can be said to 'run' in a 5' to 3' direction, but always runs in the 'opposite' direction to its counterpart.

Therefore, at the origin of replication, the two free ends of the DNA strands also run in opposite directions. One of these strands is able to grow in a 5' to 3' direction away from the origin of replication, towards the replication fork. This is the leading strand. The opposite strand needs to grow in a direction towards the origin of replication, and away from the replication fork, in order to meet the rule of being built in a 5' to 3' direction. This is the lagging strand. In this way, the mechanism replicates both DNA strands at once: the lead strand continuously towards the replication fork, and the lagging strand bit by bit, in short sequences towards the origin of replication.

3 Making new cells: mitosis and meiosis

3.1 Cell division: mitosis

For living things to grow, they must make new cells. They do this by dividing in a precise way. For the new cells to be the same as the old ones – and for them to be able to divide in turn – they must contain the exact and full DNA instructions for that organism. The process of cells dividing is called mitosis (*my-TOE-sis*) (Figure 3.1).

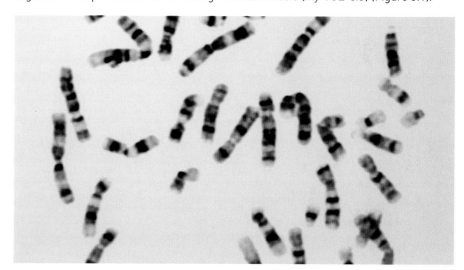

Figure 3.1 These are chromosomes in the metaphase of mitosis. At this stage they have already replicated and are ready to divide. (Image by permission of MetaSystems GmbH)

Mitosis is an efficient way of ensuring that new cells have the right genetic information (see pages 24–25). Every dividing cell replicates its full set of chromosomes, so that the two new cells it becomes contain the same DNA instructions as the old cell. All cells with nuclei are created in this way. A different method, called meiosis (*my-OH-sis*) is used to make gametes – for example, sperm and egg in humans – which need to have half the number of chromosomes.

3.2 Special cell division to produce gametes: meiosis

Meiosis (see pages 26–29) achieves two purposes very useful to the reproduction and future generations of diploid species (see page 6). The first is that meiosis reduces the number of chromosomes, so that the gamete (egg or sperm in humans) has only one

How mitosis works

Prophase

During the early part of mitosis, the DNA in the chromosomes has already replicated and the chromosomes begin to get thicker. They will become thick enough to see under an ordinary microscope. In mitosis, the cell no longer has the nucleus as a separate structure.

Every chromosome now looks like two thick threads, joined to each other at one point to make a wobbly X shape. Each of the two threads contains an identical DNA molecule. At this stage, the two identical DNA molecules in the chromosome each have one recently made 'new' strand and one 'old' strand from the DNA molecule that was in the chromosome before DNA replication and cell division started.

To get some idea of how chromosomes might appear to thicken, take a piece of wool or string at least 50 cm long, and hold one end firmly. Now start to twist the other end (ideally in the direction of its original twist), without letting either end go. As you twist, the string or wool will form coils, and then these coils will form coiled coils – which will be much thicker across and much shorter-looking than the original piece of string or wool.

The early part of mitosis is called prophase. The point at which the two chromosome threads (or sister chromatids) meet is called the centromere. The position of the centromere on the chromosome depends on the chromosome number. For instance, in chromosome 1, the largest human chromosome, the centromere is midway along the chromosome; in chromosome 21, the smallest of the human chromosomes, the centromere is near one end.

Metaphase

The next part of mitosis is called metaphase. The chromosomes have all thickened by now, and they move to the middle of the cell.

A slice taken carefully through the middle of the cell would cut through the centromere of each chromosome – because the centromeres are lined up on the 'equator' of the cell. When scientists study chromosomes in the laboratory, searching for changes in genetic material, they usually do so during the metaphase of cell division. The study of chromosomes under the microscope is called cytogenetics (Figure 3.2).

Anaphase

The chromosomes divide lengthways through the join – each X shape divides into a > shape and a < shape. Each of these new chromosomes contains the complete genetic material of its original parent as DNA, with one original, parental strand and one recently made (replicated) new strand forming its double helix. The two new daughter chromosomes move to opposite ends of the cell. The chromosomes separate through the centromere and the daughter chromosomes move apart.

chromosome from each pair the parent had. This means that when two gametes fuse to make the zygote (the first cell of the next generation), the zygote gets the complete diploid number of chromosomes. The second major purpose is that meiosis exchanges parts of the grandmother's chromosome contribution with parts of the grandfather's chromosome contribution in each chromosome pair (this is called recombination).

In mitosis, chromosomes simply make duplicates of themselves before the cell divides. But in meiosis the replicated chromosomes line up with their pairs, swap bits

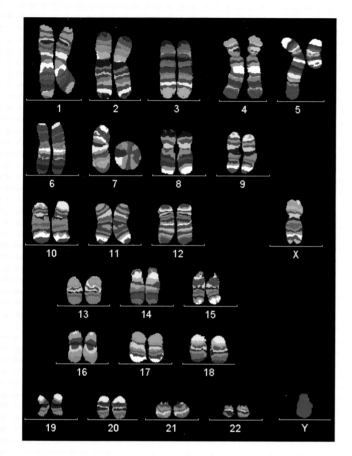

Telophase

When the cell divides, it pinches itself into two between the two groups of chromosomes – rather like squeezing a sausage balloon in the middle, except that when the sides meet they join up, so the two ends can separate. A new nucleus forms including the chromosomes of each daughter cell, and cell division is complete.

Figure 3.2 Modern techniques of chromosome analysis include this type of approach, in which specific regions of each chromosome are 'painted' with a different colour of fluorescent dye, so each looks different under the microscope. This allows cytogeneticists to tell whether the contents of chromosomes have been changed (e.g. translocations). Such changes may pass unnoticed within a person, if no chromosomal material is lost, or may cause problems to the next generation, as some material may become lost or added in meiosis, causing problems for the baby (see pages 28–29, Chapter 5). The chromosomes shown here have been 'cut and pasted' from an enhanced microscope image of a metaphase from a human male. He would be expected to have one X chromosome and one Y chromosome, and 22 more matched pairs. Can you tell whether he does? (Image by permission of MetaSystems GmbH. Chromosomes labelled with the MetaSystems mBAND method using the XCyte probe kit)

of DNA with each other, and split apart again. Then, the cell contents divide twice, in processes called the first and second meiotic divisions. The first time the cell contents divide, the cell separates the replicated chromosomes of each pair into two sets, each containing one replicated chromosome from each original pair (so that there is now half the original number of chromosomes). But now, the chromosomes are all different from the ones in the original cell, because of the way segments of DNA have been swapped. The second time the cell contents divide, the process is very much like the similar stage in mitosis. The result is a total of four cells, each containing just one chromosome from each chromosome pair. These four cells are gametes.

How meiosis works

Imagine a chromosome pair which has genes A to Z arranged along the length of each chromosome in order. An adult human has half his or her chromosomes from the mother and half from the father. Consider just one such pair of chromosomes in a man, in which one chromosome in the pair comes from his mother (shown in red) and one comes from his father (shown in blue):

ABCDEFGHIJKLMNOPQRSTUVWXYZ

and

ABCDEFGHIJKLMNOPQRSTUVWXYZ

The DNA replicates, in the same way as in mitosis, so the two replicated chromosomes are more like:

```
ABC                        VWXYZ
   DEF              PQRSTU
      GHIJ  LMNO
              K
              K
      HIJ  LMN
   DEFG           OPQRS
ABC                  TUVWXYZ
```

and

```
ABC                        VWXYZ
   DEF              PQRSTU
      GHIJ  LMNO
              K
              K
      HIJ  LMN
   DEFG           OPQRS
ABC                  TUVWXYZ
```

3.3 Why siblings can vary so much

Which chromosome from each pair in meiosis goes into the gamete is determined at random, so there is a 1 in 2 chance that any offspring of a particular pair of parents will have any particular version (or allele) of a gene present in either parent.

Meiosis explains why brothers and sisters in one family can vary so much, and why members of the whole human population can vary so much (Figure 3.3). In all 23 chromosome pairs, a human baby receives one chromosome of the pair from its mother and the other chromosome of the pair from its father. The baby's parents received their chromosomes from their parents, who are the baby's grandparents. All the genes the baby gets from its mother came from her parents, and they include the full set of genes that are meant to be on that chromosome, in the right order, but with some exchange of the baby's grandmother's and grandfather's genes. The same exchange process takes place in the baby's father's genes, so the baby will get new

Each chromosome is present in two identical copies, joined at the centromere (here shown as K). At the beginning of meiosis, the two chromosomes of each pair line up precisely beside each other, matching like for like, which we can imagine like this:

And then comes the exchange of DNA between the chromosomes of a pair, which is called 'crossing over': the two aligned, matched DNA 'arms' of the chromosome pair cross over, so some of the DNA swaps places with the matching DNA from the other chromosome of the same pair:

```
ABC                 VWXYZ
ABC                 VWXYZ
    DEF         PQRSTU
    DEF         PQRSTU
      GHIJ LMNO
      GHIJ LMNO
          K
          K
          K
          K
        HIJ LMN
        HIJ LMN
      DEFG    OPQRS
      DEFG    OPQRS
ABC              TUVWXYZ
ABC              TUVWXYZ
```

```
ABC                 VWXYZ
ABC                 VWXYZ
    DEF         PQRSTU
    DEF         PQRSTU
      GHIJ LMNO
      GHIJ LMNO
          K
          K
          K
          K
        HIJ LMN
        HIJ LMN
      DEFG    OPQRS
      DEFG    OPQRS
ABC              TUVWXYZ
ABC              TUVWXYZ
```

Figure 3.3 Five people of varying heights. Height is just one of the many characteristics that can vary hugely over the whole human population. (Science & Society Picture Library)

The result is that pieces of the man's mother's and father's chromosomes are swapped, very precisely.

After the crossing-over process, the two replicated chromosomes of each pair separate from each other through the centromere, keeping with them any exchanged DNA. This is called the first meiotic division. Notice how this is different compared to mitosis. In our example, the replicated chromosome pair would separate into:

```
ABC                          VWXYZ
   DEF              PQRSTU
      GHIJ  LMNO
           K
           K
        HIJ  LMN
     DEFG          OPQRS
ABC                     TUVWXYZ
```

and

```
ABC                          VWXYZ
   DEF              PQRSTU
      GHIJ  LMNO
           K
           K
        HIJ  LMN
     DEFG          OPQRS
ABC                     TUVWXYZ
```

These separated pieces move to opposite ends of the cell. Each end of the cell now has one set of replicated chromosomes instead of two.

The next step is the second meiotic division, which happens in just the same way as mitosis from metaphase onwards, except that it divides only one chromosome from each pair instead of two. So at one end of the cell:

```
ABC                          VWXYZ
   DEF              PQRSTU
      GHIJ  LMNO
           K
           K
        HIJ  LMN
     DEFG          OPQRS
ABC                     TUVWXYZ
```

divides into

```
ABC                          VWXYZ
   DEF              PQRSTU
      GHIJ  LMNO
           K
```

and

```
           K
        HIJ  LMN
     DEFG          OPQRS
ABC                     TUVWXYZ
```

combinations of its father's parents' genes too. Each gamete contains one or the other chromosome of each pair, at random.

3.4 Mistakes at meiosis

Occasionally in meiosis, a gamete may end up with an extra chromosome, if a chromosome has stayed with its partner instead of separating. This is called nondisjunction,

And at the other end of the cell:

```
ABC                    VWXYZ
   DEF            PQRSTU
     GHIJ LMNO
           K
           K
        HIJ LMN
    DEFG          OPQRS
ABC                    TUVWXYZ
```

divides into

```
ABC                    VWXYZ
   DEF            PQRSTU
     GHIJ LMNO
           K
```

and

```
           K
        HIJ LMN
    DEFG          OPQRS
ABC                    TUVWXYZ
```

So from one diploid cell containing two sets of chromosomes like:

ABCDEFGHIJKLMNOPQRSTUVWXYZ

and

ABCDEFGHIJKLMNOPQRSTUVWXYZ

meiosis can make four haploid gametes, each with one chromosome from each pair and one full set of chromosomes:

ABCDEFGHIJKLMNOPQRSTUVWXYZ

ABCDEFGHIJKLMNOPQRSTUVWXYZ

ABCDEFGHIJKLMNOPQRSTUVWXYZ

ABCDEFGHIJKLMNOPQRSTUVWXYZ

The same process takes place in every one of a man's chromosome pairs during meiosis. The same process also takes places in the making of gametes in meiosis in a woman, exchanging matching DNA between each pair of chromosomes she has. This means that, for each chromosome, the genes that go into an egg will be partly from the woman's mother and partly from her father.

and it can happen at either the first meiotic division or the second. The result could be a zygote which has one more or one fewer chromosome than we would expect. Sometimes this has no detectable effect, but in other cases it can stop the zygote from developing. In humans, there are only a few chromosomes that can be present in an extra copy without affecting the survival of the baby in the womb.

4 Inheritance

4.1 Are all characteristics inherited in the same way?

We, and all other diploid organisms that reproduce sexually, inherit half our genes from each parent. But what determines how we actually turn out? How exactly do we inherit our physical features and other characteristics (Figure 4.1), such as being likely to develop different diseases? Why are we not all the same? Most genes can be thought of as being 'available' in several versions, called alleles. Many of our characteristics are decided by which alleles we inherit.

The way we inherit characteristics depends, in part, on whether the genes are on the sex chromosomes or the other chromosomes. Inheritance based on the sex chromosomes is called 'sex-linked' or 'X-linked' inheritance, meaning 'on the X chromosome'. Inheritance based on the other chromosomes is called autosomal inheritance, as all the other chromosomes are known as autosomes.

The usual ways we inherit our characteristics fall into four broad categories:

- autosomal dominant
- autosomal recessive
- sex-linked (or X-linked)
- and polygenic.

Figure 4.1 Can you roll your tongue like this? In humans, one gene controls the ability to do this. Test your family and friends. (Science & Society Picture Library)

Factors from our surroundings and environment, as well as our genes, influence how we turn out.

In autosomal *recessive* inheritance, *both* copies or alleles of an autosomal gene must encode the same characteristic for that characteristic to show up in the living organism. One copy of the allele for that characteristic does not alter the individual enough to cause disease or a change that can be seen.

In autosomal *dominant* inheritance, just *one* copy or allele of an autosomal gene encoding a particular characteristic is enough for the characteristic to show. One copy of the gene changes the individual enough to cause disease or a change that can be seen.

Eye colour in humans is an example, familiar from everyday life, which includes both kinds of inheritance (Figure 4.2). Generally, people with brown eyes can either have two copies of a 'brown-eye' gene, or can have one 'brown-eye' allele (or version) and one 'blue-eye' allele of the gene. In either case these people's eyes will appear brown.

Figure 4.2 Two brown-eyed parents and their blue-eyed children. (Science & Society Picture Library)

People with blue eyes must have two 'blue-eye' genes. We cannot tell by looking at brown-eyed people, how many brown-eye genes they each have. However, when two brown-eyed people have a blue-eyed child, we know that the brown-eyed parents must each have a blue-eye gene as well as their brown-eye gene. The brown-eye gene is said to be dominant to the blue-eye gene, so brown eye colour is inherited as an autosomal dominant characteristic and blue eye colour as an autosomal recessive characteristic. This is not quite the full story of eye colour, though.

In this way, a recessive gene may pass through many generations undetected before it 'meets its match' – another recessive gene – in a child.

4.2 Autosomal recessive inheritance

Cystic fibrosis is an example of an autosomal recessive disease, which affects around 1 in 2000 babies born to parents whose ancestors came from northern Europe. The gene that determines whether we have cystic fibrosis is known as the CFTR gene.

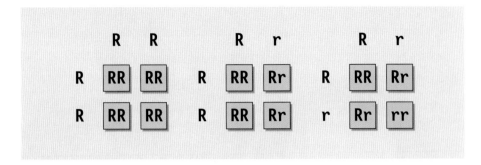

Figure 4.3 Inheritance of the CFTR gene (cystic fibrosis transmembrane conductance regulator), where R represents the dominant 'normal' allele and r is the recessive allele that causes cystic fibrosis. One parent's genes are shown across the top of each set of four squares, the other parent's genes are shown to the left of each set of four squares. The possible expected genotypes of the children are given within the squares themselves. In the left-hand set, both parents have genotype RR, so their children can only inherit genotype RR. In the middle set, one parent has genotype Rr and the other RR, so their children will have either RR or Rr, with equal likelihood. When both parents have genotype Rr, as in the right-hand set, 1 in 4 of their children will inherit genotype RR, 1 in 2 will inherit Rr and 1 in 4 will inherit the rr genotype.

We can use diagrams, showing the genes of some parents from northern Europe, to predict the likelihood of them having an affected child (Figure 4.3). Suppose that R is the dominant, typical allele of the CFTR gene – which does not cause cystic fibrosis – and that r is a recessive allele, which includes a mistake in the CFTR gene that causes cystic fibrosis when it is present in two copies. Someone with cystic fibrosis will have alleles r and r. This is described as genotype rr – geneticists use the term genotype to refer to pairs of alleles of a gene, or to alleles from the pair of genes at matching positions on a pair of chromosomes. Someone with no r allele will have genotype RR. People who have one copy of the recessive allele will not have cystic fibrosis, and will have genotype Rr. As we have seen, parents give their children either allele with equal probability.

Typically, the parents of a child with cystic fibrosis are not affected themselves, but their CFTR genes include one dominant, typical allele and one recessive allele which has a cystic-fibrosis-causing mutation. For each child they have, the random processes of meiosis decide whether that child gets the dominant or recessive allele. Thus, on average, only 1 in 4 of their children will have cystic fibrosis (or for each child there is a 1 in 4 chance that he or she will have cystic fibrosis). About 1 in 22 people of northern European ancestry has genotype Rr.

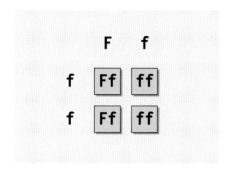

Figure 4.4 Inheritance of the LDLR gene, which can cause excess cholesterol in the blood, a condition known as hyperchol-esterolaemia. The outcome shown here results when one parent carries one dominant altered LDLR allele (F) and one recessive typical allele (f), and the other parent carries two recessive typical alleles.

4.3 Autosomal dominant inheritance

Other human characteristics and diseases show autosomal dominant inheritance. For example, some people have a faulty gene that causes them to have too much cholesterol in their blood, a condition called familial hypercholesterolaemia (FH). People who suffer from FH inherit a single copy of an altered, dominant allele of the LDLR (low-density lipoprotein receptor) gene, a gene which encodes a protein involved in the body's handling of cholesterol. This causes them to have excessive levels of LDL-cholesterol by the time they are adults. People with high levels of LDL-cholesterol are at greater risk of having a heart attack, so it is important to detect this disorder as soon as possible, so that people affected can be helped to lower their cholesterol level.

In this case, the typical 'unaltered' gene is recessive to the dominant FH gene. Around 1 in 500 people in the UK has the altered LDLR allele which causes FH. These people will also have a copy of the typical, unaltered allele, and each of their children will inherit one or other of these alleles from them, with equal probability. Thus, assuming their partner does not have FH, 1 in 2 of their children, on average, will inherit the altered gene and the risk of high LDL-cholesterol. You can see the outcome in Figure 4.4.

In general, autosomal characteristics (which cause conditions such as cystic fibrosis and FH, as we have seen) affect males and females with equal probability.

'Autosomal dominant' and 'autosomal recessive' are names given to the way the gene affects the organism as a whole. At the DNA level, both versions of every autosomal gene are present and both may be transcribed and translated. It is said that alleles of a gene are 'co-dominant' at the biochemical level. When a typical allele of a gene is dominant in terms of the features of the organism, that allele of the gene must instruct the production of protein sufficient in quality and quantity to supply the normal functions of the body. Only when there are two copies of the recessive allele do its effects show up at a whole-organism level.

4.4 Sex-linked recessive inheritance

In sex-linked (X-linked) inheritance, the special features of the sex chromosomes affect the way that characteristics are inherited. This is typical of disorders like red-green 'colour blindness', and diseases like haemophilia A, in which, typically, only men are affected. Men have only one X chromosome, which they receive from their mothers. The mother of an affected male may have the altered gene on her X chromosome, but usually shows no obvious consequences of having it, because she has another

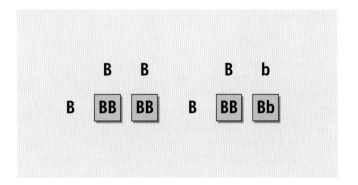

Figure 4.5 *Inheritance of the colour-blindness gene by daughters. The typical allele on the X chromosome is represented by B, while the colour-blindness allele is represented by b. The mother's X-chromosome alleles are shown along the top of the squares and the father's single X chromosome is shown to the left. In the left-hand group the mother carries the genotype BB, so her daughters can only inherit the genotype BB and will therefore all be typical. In the right-hand group, the mother carries the genotype Bb, so although 1 in 2 of her daughters will* **carry** *the b gene, they will not be colour blind.*

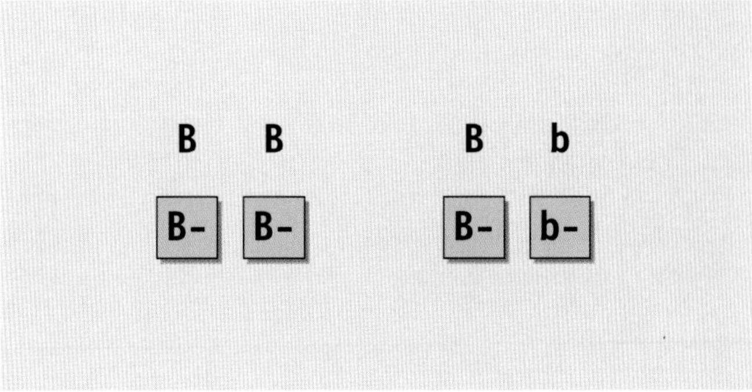

Figure 4.6 *Inheritance of the colour-blindness gene by sons. The mother's X-chromosome alleles are again shown along the top of the squares, but this time no alleles are shown for the father, as he does not give an X chromosome to his sons. In the left-hand set the mother carries the genotype BB, so her sons can only inherit the B allele and will therefore all be typical. In the right-hand set, the mother carries the genotype Bb, so 1 in 2 of her sons will inherit the b gene and will be colour blind.*

X-chromosome gene without the alteration. Thus, in a woman who carries an altered X-chromosome gene, the alteration behaves as though it were recessive, 'masked' by the effect of the gene from her other, typical X-chromosome gene.

In the example of colour blindness, a woman with normal colour vision may carry an X chromosome with the red-green colour-blindness allele on it. On average, half her sons will receive her typical X chromosome and have normal colour vision, while half her sons receive the X chromosome with the colour-blindness allele. Any son receiving an X chromosome with the colour-blindness allele will be colour blind; any son receiving the other X will not. As for her daughters, all will have normal colour vision (provided their father did), but half will carry the X with the colour-blindness allele. Again, we can see this is so with the help of diagrams (Figures 4.5 and 4.6). Some dominant and X-linked disorders have a tendency to arise as the result of new mutation (see Chapter 5), so there may not always be evidence of the child's mutation in either parent's genes. Sex-linked inheritance must also be responsible for many of the subtler differences we observe between men and women (Figure 4.7).

Figure 4.7 While many of the assumptions about the roles of men and women in Western society and the jobs that they can do have broken down over the years, there are some differences between men and women's behaviour that we cannot deny. For example, tests of men's and women's ability to solve puzzles have shown that women are better at detecting subtle hints and details, and visual memory, while men are better at seeing things in 3D, and being able to imagine how things rotate. Also, children as young as 2 usually know whether they are male or female, and can demonstrate this by choosing the kinds of toys we normally associate with boys or with girls, such as the ones shown here. Do these differences occur because of men's and women's genes, or are they a result of their upbringing? (Science & Society Picture Library)

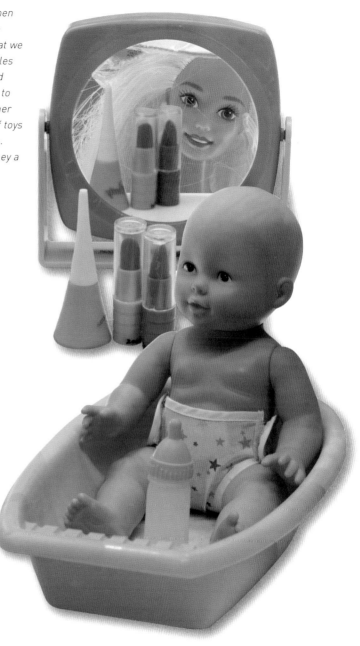

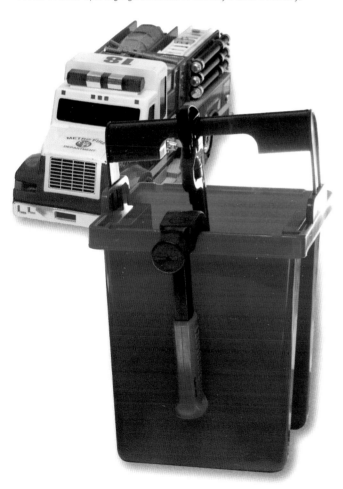

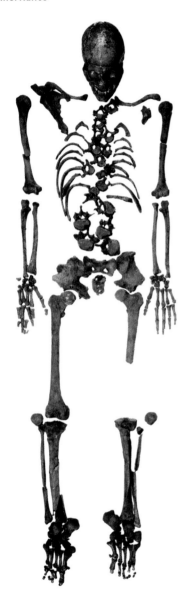

Figure 4.8 The 2000-year-old skeleton of a man found in Bleadon, Somerset. (Science & Society Picture Library)

4.5 Polygenic characteristics

Many characteristics of organisms are continuous. This means that, instead of falling into distinct categories (like eye colour in humans), the characteristics vary over a range. Examples of continuous characteristics include adult height and weight.

Most continuous characteristics and most common diseases (such as diabetes and coronary artery disease, which causes angina and heart attacks) are caused by several factors, both genetic and environmental. The genetic contribution is said to be polygenic, because a number of different genes act together, adding or overlapping their individual effects, to cause the final outcome. Many scientists are working to identify the genes that contribute most to the risks of common adult disorders. If these genes, and their risk-contributing alleles, are not the same in all populations around the world, then designing genetic tests for the all the different ethnic groups in the world will be much more difficult.

It is much easier to find the rare, single genes which make a big difference – to, say, the level of cholesterol in an adult – than it is to identify the many genes which may each contribute a tiny fraction of that effect in the general population. There are two types of difficulties geneticists have to overcome in this kind of work. The first is the biological problem of identifying the genes themselves. The second is the mathematical and statistical problem of working out whether the supposed influence of a particular gene on a particular characteristic is important for the population.

For example, although 1 in 5 British men will have a heart attack, only 1 in 500 people has familial hypercholesterolaemia. The gene that causes FH is very important to the families who have FH, but it does not explain the risk in the 99 other people in every 500 who will have heart disease. There is plenty more work for geneticists to do in understanding how genes contribute to the risks of different diseases in humans.

4.6 Mitochondrial DNA

All cell nuclei contain DNA, but not all DNA is in the cell nuclei. Interestingly, small structures called mitochondria (my-toe-CON-dree-ah), within the outer part of the cell, called the cytoplasm, also contain DNA. This type of DNA is not found in the chromosomes in the nucleus, but mitochondrial DNA is essential to life, and mitochondrial genes encode some essential proteins. Mitochondrial genes are encoded from both strands of the DNA. Mutations in mitochondrial DNA can cause disease.

Figure 4.9 When researchers studied Bleadon Man's remains, they concluded that he was 1.68 metres tall, and that he was likelier to be male because he had narrower hipbones. Further study of his bones suggested that he was right-handed and that he suffered from arthritis. (Science & Society Picture Library)

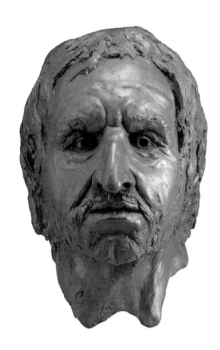

Figure 4.10 A reconstruction of what Bleadon Man's face may have looked like, made at Manchester University. (Science & Society Picture Library)

People inherit mitochondrial DNA almost entirely from their mother. This key feature of mitochondrial DNA enables scientists to trace the ancestry of organisms, because the mitochondrial DNA remains almost unchanged as it is passed down through the generations. As we have seen, meiosis mixes up the DNA contributed by grandparents to each offspring – so the further back we go in a family tree, the harder it is to work out where these bits of DNA came from. But because mitochondrial DNA doesn't get mixed up like this during reproduction, there is a trail of copied DNA into the past that scientists can follow more easily.

Comparing sequences of mitochondrial DNA has allowed us to make many discoveries. One of these is the case of 'Bleadon Man'. Scientists studied the 2000-year-old remains of a man found in Somerset, in southwest England, and found that he is related to several people living in Somerset today (Figures 4.8–4.14).

Inheritance

Figures 4.11–4.14 Guy Gibb, Doris Gould, Ruth James and David Durston, four modern-day relatives of Bleadon Man, found by studying the mitochondrial DNA of people living in Bleadon today. Can you spot a family resemblance? (Science & Society Picture Library)

Scientists studying mitochondrial DNA have also come up with theories about where on the planet the earliest humans lived and the minimum number of women there must have been to generate the diversity in mitochondrial DNA we have today. They have been able to do this because mitochondrial DNA does change, but only very gradually over a long period. It changes because of mutation.

5 Mistakes in DNA: mutation

5.1 What are mutations?

The way DNA replicates is very well organised, but there are 3,000,000,000 base pairs to copy before each cell division, so it is not surprising that mistakes sometimes happen in the precise sequence of DNA. These mistakes are called mutations, and they happen all the time, in every dividing cell. But the body is prepared for this. For example, humans have evolved a complex mechanism for correcting mistakes in DNA – there are 'teams' of proteins whose specific 'job' is DNA repair. Despite this, some mutations do escape the DNA repair process. These mutations are then copied by DNA replication and are passed on with the rest of the replicated DNA: now they are permanent (Figure 5.1).

5.2 Spelling mistakes

Mutations only matter to the organism if they affect something the DNA needs to do. They are like spelling mistakes in the DNA instruction manual, and can alter any of the functions of DNA.

Sum speling misteaks maik thyngz hardah to reed,

but may still allow the sense of the words to be understood. Other spelling mistakes may totally

grindgle xvt znirglfyjbn

into meaningless nonsense. Others still may

change only one worbd,

so most of the meaning is still clear. Similar things can happen with DNA.

5.3 Mutation and the genetic code

Think of the DNA message as being written in three-letter words in the genetic code. The first type of mutation we can think about is a simple change of one letter in a code word, which may change the meaning of the message. There are three possible results (see separate panel, Figures 5.2 and 5.3). Changing one base pair in a gene may:
- change an amino-acid
- introduce a 'stop' signal so that a protein chain is cut short
- or have no effect on the protein at all.

Figure 5.1 Instead of the rich, bright colours we come to expect from peacock feathers, here we see only white. This peacock inherited mutations in the genes that encode the proteins that make the colour in peacock feathers. (Science & Society Picture Library)

Changing one base pair in a gene: three possible outcomes

Change an amino acid

For example, a change in the DNA sequence on one strand (the sense strand) from

GCA to **CCA** means the sequence on the opposite (complementary) strand changes from

CGT to **GGT**, and the matching mRNA sequence (see page 20) produced by transcription changes from

GCA to **CCA**.

As a result, the amino-acid at the corresponding place in the protein changes from alanine to proline, a change that might affect the shape of the protein formed, and therefore its function. Think of a conventional house key: bending the uncut flat part of the key will not stop the cut part opening the lock, but bending the cut part makes the key useless. The presence of a proline suggests a change in the angle of the chain of amino-acids being formed, which alters the 3D structure of the protein, potentially affecting its function.

The effect on the protein's function of changing an amino-acid depends on the chemical and three-dimensional structure of the amino-acids concerned, and their role in that protein. The effect of changing a single amino-acid varies: it might have no detectable effect, or it could cause the protein's function to be totally lost. One important example is the cause of a condition called sickle-cell anaemia (Figures 5.2 and 5.3). The beta-globin chains in haemoglobin (the protein that gives red blood cells their colour) are encoded by the beta-globin gene. Each beta-globin chain has 146 amino-acids in it. If both copies of a person's beta-globin gene contain a change to a particular DNA base pair, that person's haemoglobin functions in a different way, and he or she is said to have sickle-cell anaemia. The cause of this is one triplet in the DNA sequence changing from GAA to GTA, so the mRNA codon changes from GAA to GUA, and the encoded amino-acid changes from glutamic acid to valine.

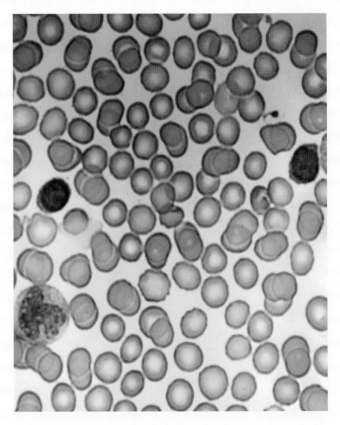

Figure 5.2 This is a magnified image of blood cells from a healthy person. Most of the circular structures you can see are red blood cells. The small, darker cell is a white blood cell and has a central nucleus and some cytoplasm around it. (Brian Bracegirdle)

People who have one beta-globin gene producing glutamic acid and one producing valine are said to have sickle-cell trait, which can be detected in the blood but may not cause disease symptoms. Blood from a person with sickle-cell trait is inhospitable to the parasite which causes malaria. In the parts of the world where malaria is common, and especially before

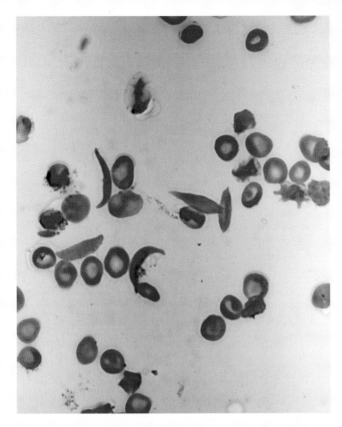

Figure 5.3 This image shows blood from a person with sickle-cell anaemia. The amino-acid change alters the behaviour of haemoglobin in the red blood cells, so that if the amount of oxygen in the blood falls, the haemoglobin molecules stick together within the red cells. This distorts the red cells into a sickle or crescent shape. You can see the characteristic sickle shape in some of the red cells in this image. The shape change alters the behaviour of the red blood cells as they circulate around the body: blood is more likely to clot inside the blood vessels and the spleen is more likely to remove the damaged red blood cells from the circulation, causing anaemia. (Brian Bracegirdle)

there were effective ways to prevent and treat malaria, people with sickle-cell trait were more likely to survive malaria than those with normal beta-globin genes or those with sickle-cell anaemia. So carrying one copy of a sickle-cell mutation gave the carrier a survival advantage.

Introduce a 'stop' signal

A change in the DNA sequence on one strand (the sense strand) from, for example,

GAG to **TAG** means the sequence on the opposite (complementary) strand changes from

CTC to **ATC**, and the matching mRNA sequence (see page 20) produced by transcription changes from

GAG to **UAG**.

This time the change is from the codon for glutamic acid to a 'stop' codon, so the protein chain formed is cut short at that point. If the new stop codon is near the end of the protein sequence, the protein may work normally; the nearer this change is to the beginning of the protein, the more likely that the protein will not work. A small chip in the end of a house key will not stop it from turning the lock, but the more of the cut part that is lost, the less likely it is that the key will work.

Have no effect on the protein at all

For example, a change in the DNA sequence on one strand (the sense strand) from

GGG to **GGT** means the sequence on the opposite (complementary) strand changes from

CCC to **CCA**, and the matching mRNA sequence produced by transcription changes from

GGG to **GGU**.

But because **GGG** and **GGU** both encode the amino-acid glycine, there is no change, and, despite the mutation, there is no consequence to the health of the organism.

Inserting and deleting base pairs

The DNA sequence

GGG TCA CCG TGT encodes

glycine-serine-proline-cysteine in the protein. Inserting an A changes this sequence to

GAG GTC ACC GTG T, which encodes

glutamine-valine-threonine-valine.

The next amino-acid will be one encoded by a DNA triplet beginning with T – an RNA codon beginning with U, which might be a 'stop' codon, because 3 of the 16 codons starting with U encode 'stop'.

Deleting a G from our sequence changes it to

GGT CAC CGT GT, which encodes

glycine-histidine-arginine-valine.

Because **GTA**, **GTC**, **GTT** and **GTG** all encode valine, whatever the next base pair is after this altered sequence, the fourth amino-acid in the sequence will be valine.

Secondly, we can think about what happens if we insert or delete a single base pair in a gene. This makes a nonsense mutation, said to alter the 'reading frame' of the gene. The transcription chemicals can only 'read' three-letter words (three-base codons), so changing the length of the message by one letter turns the outcome into nonsense. Compare

dog bit cat but man ate hat

dog itc atb utm ana teh at   (one letter deleted)

dog ebi tca tbu tma nat eha t   (one letter inserted)

Only the first of these makes sense in English. The letters are still there, but the division into words, the reading frame, has been changed. At the DNA level, this leads to the encoding of completely different amino acids (see separate panel).

The nearer a single insertion or deletion of a base pair is to the start of a gene, the greater the effects of the 'nonsense' amino-acids will be on the protein's function. Adding or subtracting two base pairs will also alter the reading frame and result in a nonsense product. Frame-shift mutations like these may result in a new stop codon, later or earlier than that in the original sequence, leading to the production of a protein that is bigger (more amino-acids) or smaller (fewer amino-acids) than normal, as well as being different in composition.

The cause of cystic fibrosis

Cystic fibrosis, a disease that affects the lungs and digestive system, occurs in people who have two defective copies of the CFTR gene. Hundreds of changes to the DNA sequence in the CFTR gene can cause cystic fibrosis. One of these changes is the deletion of the codon for the amino-acid phenylalanine, the 508th amino-acid in the finished protein. About 1 in 22 healthy people whose ancestors came from northern Europe carry one copy of the CFTR gene with a mutation (Figure 5.4). Carriers of a mutation in the CFTR gene are thought to have had a survival advantage in the past, as we have seen with the sickle-cell trait. Scientists have proposed various theories as to what this advantage may have been. Some favour the intuitively attractive idea that people who carried a cystic fibrosis mutation may have been less likely to die from fluid and chemical loss if they contracted diseases such as cholera.

Figure 5.4 It is likely that 1 in 22 of these people has one CFTR gene with a mutation. (11010010, stock.d2.hu)

Inserting or deleting three base pairs will not alter the reading frame, but will respectively add or subtract an amino-acid. How important this is to the protein's function depends on the position of the change in the finished protein and its effect on the protein's shape. Going back to our key analogy, changing the colour of the key will not stop it working, nor will cutting a 'pie slice' from the handle part. But shortening or lengthening the cut part will stop it fitting the lock.

The faulty version of the CFTR gene, which we encountered earlier (see pages 31–32), is an important example of a mutation arising from the deletion of three base pairs (see separate panel, Figure 5.4).

Even if an amino-acid is changed, some base changes in the coding part of genes have no obvious damaging effect on encoded proteins, because some amino-acid changes leave the essential functions of their proteins unchanged. Mutations causing these types of changes do not affect the health of the organism, but contribute to variability among individuals within a species that can be detected (Figure 5.5).

Figure 5.5 These butterflies belong to the same species. The different colours and patterns on their wings are controlled by just a few genes. (Science & Society Picture Library)

Figure 5.6 This cat has seven toes on each paw, as a result of a mutation. The same types of genes also govern the number of fingers and toes in humans. (Science & Society Picture Library)

Mutations can affect DNA sequences that regulate the function of genes, so that a gene may make too much or not enough protein, or the right protein at the wrong time, or not enough protein at a crucial stage of development, for that stage to proceed normally (Figures 5.6 and 5.7).

In the cell in which a mutation happens, the protein may lose its function completely. The effect on the living organism of total loss of protein function depends on how important the protein is to the cell, in location and timing, and how important the cell is to the organism as a whole. A mutation in a gene causing damage to a human

45

heart-muscle protein will be less important if it happens in the DNA of a cell in the skin of the face, than if it happens in the heart muscle. A mutation in a gene which decides the order of formation of parts of the eye of a developing fly may not matter if it happens in the eye of an adult fly.

The effect of mutations on the living organism is important to the descendants of that organism – and potentially, to the species – if the mutation is inherited. Mutations which are going to be inherited must be present in the gametes (egg or sperm in humans) which will form the zygote, the first cell of the offspring organism.

5.4 Tumours

Occasionally, a mutation in a single cell can allow it to start growing out of control. This can cause a tumour; if the effect of the mutation is sufficient to overcome the body's usual limits on growth and overgrowth, the tumour may become cancerous and spread to other parts of the body. A mutation need not necessarily be inherited to result in disease or death. All living organisms are exposed to causes of mutation every day, including ultraviolet light, cosmic rays and other forms of radiation, and toxic chemicals. Extreme doses of radioactivity can cause severe DNA damage, extensive mutation and death in any living organism.

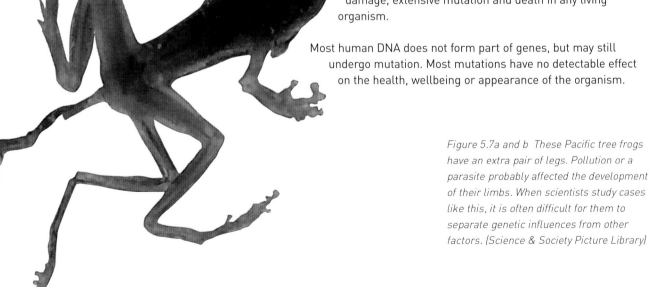

Most human DNA does not form part of genes, but may still undergo mutation. Most mutations have no detectable effect on the health, wellbeing or appearance of the organism.

Figure 5.7a and b These Pacific tree frogs have an extra pair of legs. Pollution or a parasite probably affected the development of their limbs. When scientists study cases like this, it is often difficult for them to separate genetic influences from other factors. (Science & Society Picture Library)

5.5 Mutation and diversity

Mutations occurring over many thousands of years have caused the differences between individuals of the same species that we see today (see separate panel, Figures 5.8–5.11). Studies of the types of differences, at DNA level, between different human populations of today have led to theories about how human populations migrated around the world. Scientists believe that the first representatives of *Homo sapiens* evolved in Africa around 200,000 years ago, and that their descendants later spread across the world.

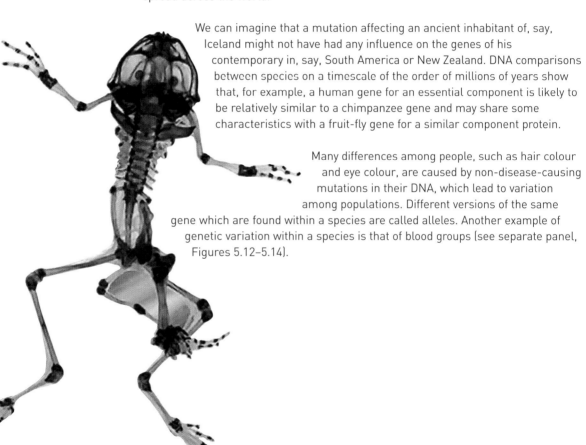

We can imagine that a mutation affecting an ancient inhabitant of, say, Iceland might not have had any influence on the genes of his contemporary in, say, South America or New Zealand. DNA comparisons between species on a timescale of the order of millions of years show that, for example, a human gene for an essential component is likely to be relatively similar to a chimpanzee gene and may share some characteristics with a fruit-fly gene for a similar component protein.

Many differences among people, such as hair colour and eye colour, are caused by non-disease-causing mutations in their DNA, which lead to variation among populations. Different versions of the same gene which are found within a species are called alleles. Another example of genetic variation within a species is that of blood groups (see separate panel, Figures 5.12–5.14).

Genetic fingerprinting

All humans have the same genes at the same places on the same chromosomes. Even so, there are so many small differences in the DNA sequence between one person and another that each person's DNA can be considered as a unique 'fingerprint'. Figure 5.8 shows a physical human fingerprint; Figure 5.9 shows a genetic 'fingerprint'. DNA 'fingerprints', which are nothing to do with fingers, were first identified in the 1980s.

Scientists have several different ways of looking at human DNA and producing genetic fingerprints. But nowadays they usually use a method that relies on a chemical process called the polymerase chain reaction (Figure 5.10). Scientists prefer this technique because even tiny amounts of DNA can be used.

First, the DNA has to be extracted from the cells (see page 14). Then the polymerase chain reaction is used to 'amplify' particular sequences from the DNA extracted. In other words, the chemical process begins with the small amount of DNA

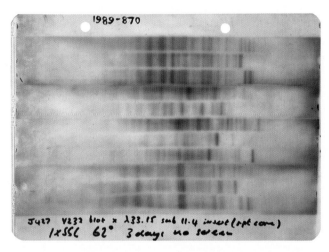

Figure 5.9 This is the first ever picture of genetic fingerprints, produced by Professor Sir Alec Jeffreys in 1984. It is an original autoradiograph, a photographic film on which the bands are made by a radioactive chemical attached to the different DNA fragments. Bands in a DNA fingerprint are inherited – half from the mother, half from the father. This allows DNA fingerprints to be used to test whether or not a particular man is the father of a child. But because of the way DNA is inherited (see Chapter 3), the bands are also so variable among people that each person's can be considered almost unique. This means that forensic scientists can use this technique to match DNA samples with the people they came from. (Science & Society Picture Library)

Figure 5.8 Every human has different fingerprints from every other human – even identical twins – so prints on the fingers are not entirely determined by genes. Fingerprints are laid down early in the development of a child in its mother's womb.

recovered from the human cells, and then multiplies the quantity of DNA so that there is more for the scientist to work with (Figure 5.11).

The next step is to make the DNA visible, using a technique called gel electrophoresis. DNA fragments have a negative electric charge, so they can be made to move by an electric current. The scientist places the DNA in oblong 'wells' on a sheet of gel, submerged in a special chemical solution. Electrodes, one positive and one negative, are immersed in the liquid.

As a result, when a current is switched on, the amplified DNA moves towards the positive electrode. Bigger DNA fragments move more slowly than smaller ones, so fragments of different sizes move different distances through the gel. After a while, the current is switched off. The scientist then uses ultraviolet light to make the DNA in the gel 'fluoresce' and therefore become visible.

Figure 5.10 Setting up a polymerase chain reaction at the Institute of Child Health, London. (Science & Society Picture Library)

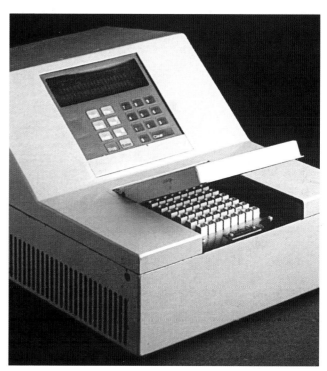

Figure 5.11 'Baby Blue', a prototype polymerase chain reaction machine, made at the Cetus Corporation, California, in 1986. The machine automates the process of making lots of DNA from a tiny starting sample. (Science & Society Picture Library)

Blood groups

Blood groups provide an example, which we take for granted, of genetic variation among human beings. The ABO blood-group system arises because there are three different versions of the same gene in human populations. The three versions of this gene account for blood group A, blood group B, and 'not-A and not-B', which is known as O. Everyone has two ABO blood-group genes, that is, two alleles of the ABO blood-group gene, which may be the same as each other or different.

The A and B genes make proteins which affect red blood cells in different ways. The O genes make protiens that do not affect red blood cells. So if blood with group A or group B is given (transfused) to someone who has group O, that person's body treats the transfused blood as an invader – it reacts against the 'foreign' red blood cells in the group A or B blood. However, blood from someone with group O has no 'changed' red blood cells, and is not rejected by someone who makes blood-group-determining proteins A or B. The situation is also complicated by D or rhesus blood typing (see Figure 5.12 and below).

Think about all the people who have blood group A. Their genes can be either

- two A genes, or
- one each of A and O.

People with blood group B can have either

- two B genes, or
- one each of B and O.

People with blood group AB must have one gene each of A and B, and people with blood group O must have two O genes. We can work out the blood-group genes of children by starting from the blood-group genes of their parents – and sometimes we can work out the blood-group genes of parents from their children.

Consider a couple whose blood has been tested, where it was found that one parent has blood group A and the other has blood group B. There are four possibilities:

- one parent carries *two* A genes, and one carries *two* B genes
- one parent carries *two* A genes, and one carries *one* B gene and *one* O gene
- one parent carries *one* A gene and *one* O gene, and one carries *two* B genes
- one parent carries *one* A gene and *one* O gene, and one carries *one* B gene and *one* O gene.

The possible blood-group genotypes in the children of these parents are shown in Figure 5.13.

The ABO system is not the only blood-group system in humans. Another important kind of blood group in medicine is the rhesus blood group, also known as the D group. It is controlled by a single gene, which decides the presence or absence of D protein in the blood. The D group also is important in transfusion: people who have D-positive, or rhesus-positive, blood can be given D-negative, or rhesus-negative, blood. But people with D-negative blood are not given D-positive blood, as they will react against the D protein. People with one or two copies of the allele that puts the protein in the blood are said to be D positive. People with no copies of the D-protein-in-the-blood allele – who therefore have two alleles for 'no D protein in the blood' – are said to be D negative.

The D blood-group system is particularly important because it can cause a problem in pregnancy. If a D-negative woman and a D-positive man have a baby, and if the baby is D positive, then a little of the baby's D protein may make its way into the mother's blood stream before the baby is born. This D protein may act rather like a vaccine in the mother, so her body prepares an immune response, ready for any future occasions when she is exposed to D protein. In a later pregnancy by a D-positive man, her body may mount an immune response against any D protein that reaches her immune system, by making her own 'anti-D protein'. If this happens, the consequences for the child are potentially serious. Fortunately, modern medicine can easily prevent this problem. Pregnant women are given a blood-group test (Figure 5.14), and their blood is also checked for anti-D

Figure 5.12 Blood groups are named according to ABO typing and D (rhesus) typing. From left to right, the blood groups of the blood in these transfusion bags are A positive, B negative, AB negative and O positive. Apart from these major blood-group types, there are also other blood-group types under genetic control. Some of these are rare variants, known as private blood groups. Rather than test every patient's genes for every possible blood-group variant, medical workers carry out a quick and easy test of compatibility. Normally, before a patient is given a blood transfusion, a sample of the blood to be transfused is 'cross-matched' against a sample of the patient's own blood. The blood itself contains the components that might 'reject' a transfusion, and the rejection reaction causes a visible change in the mixture. If there is no reaction, it should be safe to transfuse the blood into that particular patient. Even so, hospital staff watch a patient carefully when a transfusion begins, in case there is some incompatibility that was not detected by the earlier tests. (Science & Society Picture Library)

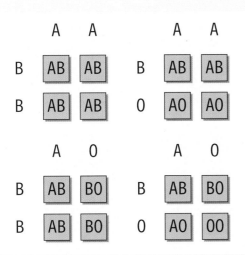

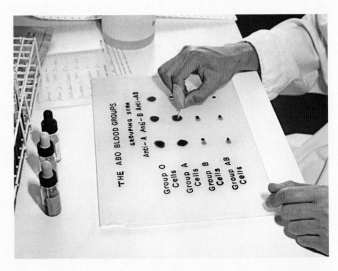

Figure 5.13 The possible ABO blood-group genotypes of the children of a couple, one of whom has blood group A and one of whom has blood group B. The first parent's two blood-group genes (alleles) are shown across the top of each set of four squares, and the second parent's two blood-group genes are shown to the left of each set of four squares. The contents of the four squares show the possible ABO blood-group genotypes in the children and hence what the results of their blood-group tests would be. In the first set, all children would have blood group AB. In the second set they are equally likely to have group AB or group A, and in the third set they are equally likely to have group AB or group B. In the fourth set, children are equally likely to have blood group AB, A, B or O.

Figure 5.14 Testing for blood groups in 1980. (Science & Society Picture Library)

People whose blood group is AB, D positive are called universal acceptors – most other people's blood can usually be given safely to them.

Blood for transfusion is usually matched for compatibility with the person receiving it.

A blood transfusion is a simple form of transplant. When organs are transplanted, the donor and recipient may be tested for a rather more sophisticated range of transplantation (histocompatibility) proteins, many of them produced on the cell surface, and controlled by a set of genes. Identical twins have the same genetic profile, so they have the same blood group as each other. Brothers and sisters or parents and their children share, on average, about half of their genetic profile and proteins. Unrelated donors and recipients may share fewer compatible proteins, but advances in modern medicine mean that treatment can overcome some of the problems associated with incompatibility and make it less likely that a recipient will reject a donated organ, even one that is apparently incompatible.

during the pregnancy. All pregnant women who are D negative are given an injection of anti-D just after childbirth, to 'mop up' any passing D protein. This prevents an immune response occurring if these women go on to have another pregnancy by a D-positive man.

People whose blood group is O, D negative are called universal donors – their blood can usually be given safely to most other people.

6 Genetic engineering

Figure 6.1 Genetically engineered mice belonging to a strain developed at the Harvard Medical School in 1988. This type of mouse was the first transgenic mammal to be granted a patent in the USA. The mice's genes were altered to make them more prone to develop cancer, thereby making them very useful in cancer research. (Science & Society Picture Library)

6.1 What is genetic engineering?

Genetic engineering is the deliberate alteration of DNA. It can refer to a wide variety of techniques used by scientists (Figure 6.1). These techniques range from merely investigating DNA function to cloning human tissue or, indeed, humans themselves. Whether we want to amend the DNA of a bacterium, a plant or a mammal, the main principles behind genetic engineering are the same – but, in practice, the techniques become much more difficult for more complex organisms.

Many people are concerned about genetic engineering: whether it is safe, beneficial or even acceptable in principle (see Chapter 7).

6.2 Cloning DNA

One fairly simple method scientists use in genetic engineering is inserting sections of human, animal or plant DNA into the DNA of 'vectors'. These vectors are based on DNA 'parasites' that replicate independently inside bacteria. The vectors will copy the extra DNA inserted into them by the scientists, along with their own DNA. Bacteria replicate a lot faster than more complicated organisms such as humans, and the vector DNAs often replicate faster still. Scientists can exploit the vectors' tendency to replicate rapidly by adding a gene to the vector that gives the bacteria resistance to antibiotics – so the bacteria have a survival advantage.

The end result of all this replication and survival is a large amount of DNA identical to the original DNA the scientists inserted into the vector. This is 'cloning' at a DNA level. It was the method scientists developed to prepare enough DNA to study in detail, and it is still used today. Cloned DNA can be adapted to make useful proteins for medical uses, such as insulin for the treatment of diabetes.

6.3 Transgenic animals

All species use their DNA in the same sort of way, even though their genes are not identical. As a result, a gene from one species can be used by the cells of another (Figure 6.2). The effects of this depend on exactly where the gene is inserted and how it is transcribed. Some of the DNA vectors mentioned above have been modified by scientists so that they contain regulatory DNA sequences near a particular site. When a gene is cloned into that site, the vector's DNA regulatory sequences actually use the inserted gene to make a required protein.

Vector DNA can be made to produce a protein from another species in bacteria. This led scientists to develop methods that allow animals to produce useful proteins. In a transgenic animal, DNA from a different species replicates within the host animal and uses the host's protein synthesis apparatus to make a particular protein. This protein

Figure 6.2 Tracy, a transgenic sheep, lived from 1990 to 1998. She produced a human protein in her milk, as she had human DNA in her own cells. It was hoped that this protein, alpha-1-anti-trypsin, would be useful in treating the symptoms of cystic fibrosis (see pages 31–32, 43). Scientists use transgenic animals for a number of reasons. For example, they can test whether replacing a particular defective gene can cure a disease. Often they use mice, which are easier to handle and breed than sheep. (Science & Society Picture Library)

will be of a precise type and amino-acid sequence, so that, for example, a human protein could be made in another species and then given to humans who lack that protein.

Among the proteins produced in humans are the blood-group proteins (see pages 50–52) and the specialist 'transplantation antigens', which are the trigger determining whether the body accepts or rejects a transplant. Genetic engineers have put human genes into pig cells, so that pig heart cells might make human transplantation antigens. This was done because scientists hoped it might become possible to use hearts from pigs, instead of hearts from humans, for transplant operations (Figure 6.3).

6.4 Genetically modified plants

Altering the genes of plants is easier than altering the genes of animals. It is also less controversial, at least at the laboratory stage. For thousands of years, people have known that plants and animals can be modified by breeding selected individuals with specific others. Genetic engineering has now made it possible to control the outcome exactly in some cases. For example, scientists have succeeded in producing disease- and pesticide-resistant crops. As a result, breeders of plants and animals no longer have to rely on the hazardous route via mutation and natural selection to obtain the results they want. The characteristics of food plants (such as colour, shelf-life, size, etc.) can be changed individually by altering the specific genes which control them, but only when scientists know which genes these are.

6.5 Cloning whole organisms

There is an enormous technological leap from cloning a piece of DNA in a vector, within a bacterial host, to cloning an entire organism (Figure 6.4). This involves placing the complete DNA of one organism into a DNA-free, primitive embryonic cell from another organism of the same species. If this primitive cell divides successfully, then, in the early stages of its multiplication, each daughter cell retains the capacity to form a whole organism. This is a known characteristic of the cells from the very earliest stages of development of an embryo. If each cell of this kind was placed into an environment suitable for embryonic development, each primitive daughter cell would give rise to a genetically identical organism – a clone (that is, a newborn animal, not a 'copy' of an adult animal).

Animal cloning is a real possibility, and has been achieved with some success, though it is clear that scientists are not yet aware of all the ways a cloned animal might be affected, throughout its life, by its being grown in this way. A clone does not have the

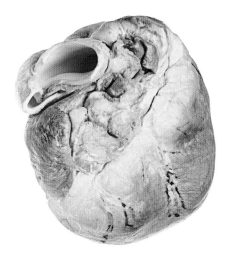

Figure 6.3 A pig heart which has had human genes engineered into it. A pig heart is the right sort of size and power for a human. Scientists have inserted human genes into pig genetic material in an attempt to make the pigs' hearts more compatible with humans. (Science & Society Picture Library)

Figure 6.4 *Some pig organs are relatively similar to human organs. Scientists hope that their work in cloning ordinary pigs like these will eventually lead to the possibility of farming pigs to supply organs for transplantation into humans. (karel, stock.d2.hu)*

experiences that the original owner of its DNA had, and has not been exposed to the same things. But it does have the accumulated damage and mutations that the DNA encountered in the cell it was in within its original 'host'. There is not (yet) any way to check that the DNA going into a clone is 'perfect'. The best-known example of cloning in Britain is Dolly the sheep (see separate panel, Figures 6.5 and 6.6).

6.6 Gene therapy

It is important to remember that having a gene for particular disease does not mean the affected person is definitely going to suffer from that disease. In many cases, environmental changes, drugs or surgery can help diminish the effects of a particular disease gene. For severe diseases, however, where these interventions are not enough to counter the effects of the genes, scientists are working towards gene therapy.

The simplest concept of gene therapy is the idea that adding a normally functioning copy of a gene to a cell with two defective copies might be able to restore normal function. There are many obstacles to successful gene therapy, but a massive worldwide scientific effort is going into it. We must hope that a build-up of discoveries, like those which preceded the discovery of the structure of DNA, will lead to the breakthroughs many families long for (see pages 61–62).

Dolly the sheep

Cloning of animals is possible because DNA from any cell contains all the instructions to build the whole organism. The birth of Dolly the cloned sheep, in 1996, caused something of a sensation in the world of science. The cell that eventually became Dolly was produced when DNA from an adult sheep's mammary gland was injected into sheep eggs which had had their DNA removed. An electrical 'shock' was then enough to 'trick' the egg and the donor DNA in it into behaving like a newly fertilised egg – so it divided and grew to form Dolly when it was implanted into the womb of another sheep.

The ends of chromosomes are called telomeres, and are made up of long series of repeats of DNA sequences. When a cell divides and the DNA replicates, it stops short of replicating quite all the telomere repeats, with the result that the telomere gets shorter. This has an interesting side effect: if a cell from an adult of a species is used to clone a new member of the same species, the clone is potentially disadvantaged from the start by having shorter telomeres than a normal newborn. It has been suggested that this might lead to premature ageing in clones – they could become 'old before their time'. Dolly the sheep died from a lung infection when she was 6 years old, which is quite young for a sheep. She had also developed arthritis. Did Dolly die prematurely because her DNA was older than her chronological age?

Figures 6.5 and 6.6 The front and back of a jumper made using wool from Dolly, a cloned sheep. The jumper was designed by Holly Wharton, who was 12 years old at the time and had won a competition sponsored by the Cystic Fibrosis Trust. Her design was knitted up using wool from Dolly's fleece. (Science & Society Picture Library)

7 Genetics and ethics

7.1 Many debates

Knowledge can bring dangers as well as benefits. Because the media like to run stories that are as dramatic as possible, they may exaggerate the likely benefits of new technological advances, or they may sensationalise the possible dangers. Different people have different opinions about which genetic technologies should be approved and which should not. Their opinions will arise from their basic beliefs – which could be religious, cultural, scientific or moral beliefs, or combinations of these. Governments decide what should and should not be allowed, and in many countries they will consult experts and members of the public, as well as politicians, to decide where to 'draw the line'.

Figure 7.1 A genetically engineered tomato plant, modified to slow down the ripening process. (Science & Society Picture Library)

Genetic engineering makes it possible to alter genes, or to add them. For example, scientists have been able to change the growing characteristics of crop plants to make the food last longer or the crops more resistant to disease (Figure 7.1). Some people feel it is wrong to 'tamper with nature' in this way, while others say that genetically engineered crops should be used to combat world poverty, or to make food cheaper or easier to produce and store.

If scientists can alter or add genes in plants, they can do the same in animals: they can produce genetically engineered animals that are fatter or thinner or meatier, or which make a protein for doctors to use in treating disease (Figure 7.2). Some people who think it is acceptable to alter the genes of plants think it is wrong to change genes in animals. Others think it is acceptable to introduce new genes into an animal, as long as they are 'fixed' so they cannot be passed on to new generations. Others think that limiting genetic modification of animals is too restrictive. Cloning bacteria seems acceptable enough – but cloning people seems to be a step too far for many of us (Figure 7.3).

7.2 Genetic tests

At first sight, the possibility of carrying out genetic tests might sound relatively uncontroversial, compared to other genetic technologies. In fact, it raises a great many ethical issues. Geneticists are no strangers to these considerations, and people working in laboratories where genetic disorders are diagnosed always consider carefully the consequences of carrying out new tests. Children in the UK are generally not tested to determine their genetic risk of developing a disorder that may only affect

Figure 7.2 Scientists produce genetically engineered animals to help them find out how genes affect behaviour. This obese mouse was produced at the Rowett Research Institute in Scotland by 'knocking out' a gene responsible for influencing the animal's appetite and feeding behaviour. (Science & Society Picture Library)

them much later in life. They can then decide for themselves whether to have a test when they are old enough.

Geneticists can now determine much more exactly who is likely to suffer from a particular inherited disease. There is a risk that people identified in this way could end up being discriminated against (in some societies they might find it difficult to get life insurance, for instance). But if geneticists could identify every genetic disorder, everyone in the world would probably turn out to be at risk of having or transmitting one of them.

Some people feel it is wrong to test unborn children for genetic diseases. Others feel that tests should be allowed only when doctors suspect a severe disorder. There are people who believe they should have the right to choose the sex, or even the appearance, of their child in advance. People are sometimes against testing of unborn children because they fear parents will use the information to decide to terminate a pregnancy. Parents may make this choice if they know their child is going to be affected by a serious disease. Some parents may seek these prenatal genetic tests so that they can prepare themselves fully for having an affected child.

Figure 7.3 Although, as far as we know, no-one has yet been able to produce a clone of a human, it remains a distinct possibility in the future. This is a 'home cloning kit' produced by the Genemsco Corporation in 1984. It was supplied to schools to promote careers in genetic science. The kit came with parent bacteria in which students could clone DNA. (Science & Society Picture Library)

In the UK, when clinical geneticists and genetic counsellors see couples and families, they try to offer them a form of counselling in which they help the patients to understand the risk information appropriate to them. They do not pressure families into following any particular course of action.

It is important to remember that, even for a dominant disorder that emerges in adult life, on average, 1 in 2 of the children of an affected person will not inherit the disease gene. For these 50 per cent of people at risk, a genetic test can reassure them and help them avoid anxiety. Others choose not to have a test, and their decision is respected too.

7.3 Who owns genetic information?

Like medical information, genetic information is treated as confidential in the UK – though with extra safeguards, because genetic information on one person has implications for that person's whole family. But what should geneticists do when someone doesn't want to share his or her genetic information with relatives? If the information was passed on, the relatives could find out they were at risk of a disease they did not know was in their family – or that they were not at risk of a disease they had considered a threat (Figure 7.4).

What do you think?

Here are some scenarios and questions to consider. There are no right or wrong answers to the questions.

Testing for disease

Would you want to be tested for your risk of suffering from a rare genetic disorder which would not affect you until the age of 60? What if it was a common disorder? What if one of your relatives had it? What if your children would be at risk? And what if your children, once they were grown up, wanted to know their risk?

Imagine that people could be tested when they are children for their risk of getting or passing on genetic diseases. This information could then be kept on file, and cross-checked against a prospective partner to see whether any children they might have would be at risk of a particular disease. What should happen to this information? Should the couple get the information and make their own choice? Who should decide who has the rights to the information? (You probably are a carrier of disease alleles for several recessive diseases. They are most likely to be rare diseases, and the chance of you having an affected child is very low unless your child's other parent also carries an allele for the same disease.)

Tests on animals

Under what circumstances is it ethical or unethical to use animals for genetic experiments that may benefit our understanding of human disease? If you have very strong feelings about this, one way or the other, think whether there are any circumstances under which you would change your view.

Crops

If a genetically modified food crop could reduce world hunger by half, at the expense of the unmodified species of crop being lost for ever, would introducing the crop be a good idea? What if the modified crop could reduce hunger by only a quarter, or 10 per cent? If not, why not?

Clones

What rights and identity would a cloned person have? What if a clone grew up to be just like the donor of its DNA – what rights should the cloned individual have? Should a cloned individual be allowed to have children? If not, why not? Who should decide the rights of a clone?

Some governments are considering whether they should sell genetic information about their populations to drug companies, to help the companies with their research. When the government of Iceland decided to store genetic data on the country's population it caused a major controversy.

7.4 Gene therapy

Gene therapy offers hope for curing some diseases, provided that some obstacles can be overcome. In the UK, scientists are not allowed to use gene therapy to alter genes that will be passed on to the next generation ('germ-line gene therapy'), only to alter genes in the organs of an existing person. This is because we do not know enough about the possible side effects of incorporating a new gene permanently into the DNA

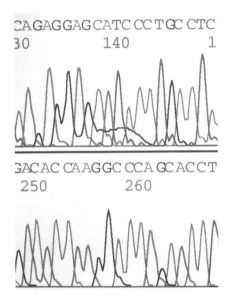

CA GAG GAG CATC CC TGC CTC
30   140   1

GAC AC CA AGGC CCA GC ACC T
250   260

Figure 7.4 In the future, health services may keep on file information about everyone's DNA. (Science & Society Picture Library)

of future generations. Some of the people who have the disorders which may be treatable by gene therapy think germ-line gene therapy is a good idea, because it could make it possible for them to have children with a normal allele instead of a disease-causing allele of the gene.

When obstacles to gene therapy are overcome, doctors might be able to use it, for example, to delay the effects of ageing (Figure 7.5). But how would society – or global resources – support an ageing, healthy population?

7.5 Genetically modified crops

If the germ-line technique is used when crops are genetically modified, the plants may be able to make genetically modified pollen, which insects can then transfer to unmodified plants in other places. Any seeds and new crops generated via this route would then be hybrids between the original and genetically modified versions of the species. This is one of the major worries of environmental campaigners who are opposed to GM crops.

Genetic engineers working for major food manufacturers have tried to respond to these worries in various ways. For instance, it is possible to make genetically modified plants that require some nutrient added to the soil for them to be able to grow. These plants' descendants would then also require that same nutrient, and would only grow on soil that had been specially prepared.

7.6 Cloning organs, animals and people

The stem cells that constitute an embryo in its earliest stages of development are capable of developing into any part of the body. Scientists think that it ought to be possible to generate stem cells containing a person's own DNA. These cells could then be used to 'grow' particular organs (such as a liver) or tissues (such as skin) to treat that person. Many scientists think this might be a reasonable application of cloning methods for humans. Do you agree? Would such an organ be safe for its recipient?

Growing a spare organ in the laboratory for someone might be a major advance over the use of organs from animals. Using organs from animals to treat disease in humans would involve genetically engineering cloned animals to generate human-compatible organs. But using animal organs in this way has some disadvantages: the differences between the donor species and humans will probably always cause difficulties, and some people might consider the practice to be unacceptable exploitation of animals.

Figure 7.5 Five people of varying ages. Parts of the body wear out as it gets older – but it might be possible to delay the effects of ageing using gene therapy. Should we be trying to do this? Would your answer be different if you were 20 or 50 years older? (Science & Society Picture Library)

Dolly the sheep's arthritis (see page 57) may be an example of the health problems, caused by premature ageing, thought likely to beset cloned animals. Or it may have arisen by chance. Scientists in some parts of the world may already be attempting to clone humans – and one company linked to a religious cult claims to have done so already. Most scientists feel the state of our knowledge is not ready to attempt cloning of humans, and many see no real need for it. Should it be permitted?

7.7 'Perfect' DNA

People sometimes talk about 'designer babies', imagining them as babies 'made to order' with physical characteristics (for example height or body shape), intelligence, health and abilities (sporting or musical ability, for instance) chosen by their parents. Not every physical characteristic is genetically determined, and for many that are, the mechanisms are not fully understood. For example, we do not yet fully understand the genetic contributions to intelligence or height, though both are likely to have a polygenic genetic component. Some people find the idea of 'designer babies' abhorrent; others think they could enable future generations to lead happier, healthier lives. What do you think?

Index